DEPARTMENT OF THE INTERIOR
UNITED STATES GEOLOGICAL SURVEY
GEORGE OTIS SMITH, DIRECTOR

BULLETIN 478

GEOLOGY AND ORE DEPOSITS NEAR LAKE CITY, COLORADO

BY

JOHN DUER IRVING

AND

HOWLAND BANCROFT

WASHINGTON
GOVERNMENT PRINTING OFFICE
1911

CONTENTS.

ILLUSTRATIONS.

INSERT.

GEOLOGY AND ORE DEPOSITS NEAR LAKE CITY, COLO.

By JOHN DUER IRVING and HOWLAND BANCROFT.

FIELD WORK.

The field work on which this report is based was begun by the United States Geological Survey in the summer of 1904. During August of that year J. D. Irving spent a month in the district examining the properties then accessible and collecting notes and specimens for study. In the summer of 1908 new developments in the region made another examination expedient, and Howland Bancroft was sent to collect additional information and to bring Mr. Irving's observations up to date.

The mines have been worked for a long time and in a desultory manner and many of them have been abandoned, so that only a small part of the ore bodies that have been worked can now be examined. The data collected are thus necessarily fragmentary and the writers' generalizations are the result of field work done under somewhat difficult and unsatisfactory conditions.

This report is written jointly by Mr. Irving and Mr. Bancroft and is based both on the original field observations and the later observations and collections. For the chapter on the geology of the region the authors are indebted to Whitman Cross, who is also largely responsible for the geologic map.

ACKNOWLEDGMENTS.

To the owners and operators of the mines in the Lake City district the cordial acknowledgment of the authors is extended for maps of workings, statistical data, opportunities to study cost sheets and assay results, and unreserved permission to enter and study all accessible underground workings.

The long and eventful history of this camp and the changes of personnel in the mining population consequent on the many alternating periods of depression and intense activity have made the collection of accurate historical information very difficult. The history of the district (pp. 12–14) has been compiled chiefly from notes furnished by Mr. J. J. Abbot, who has been long and intimately associated with the region and whose courteous aid is highly appreciated by the writers. These notes would be inserted in full as a section by Mr. Abbot if the limitations of space permitted.

GEOGRAPHY.

LOCATION AND GENERAL FEATURES.

The Lake City quadrangle is in southwestern Colorado, between longitude 107° 15′ and 107° 30′ W., and latitude 38° and 38° 15′ N., adjoining the Ouray quadrangle on the west. (See fig. 1.) It covers about 235.7 square miles and contains some of the most rugged

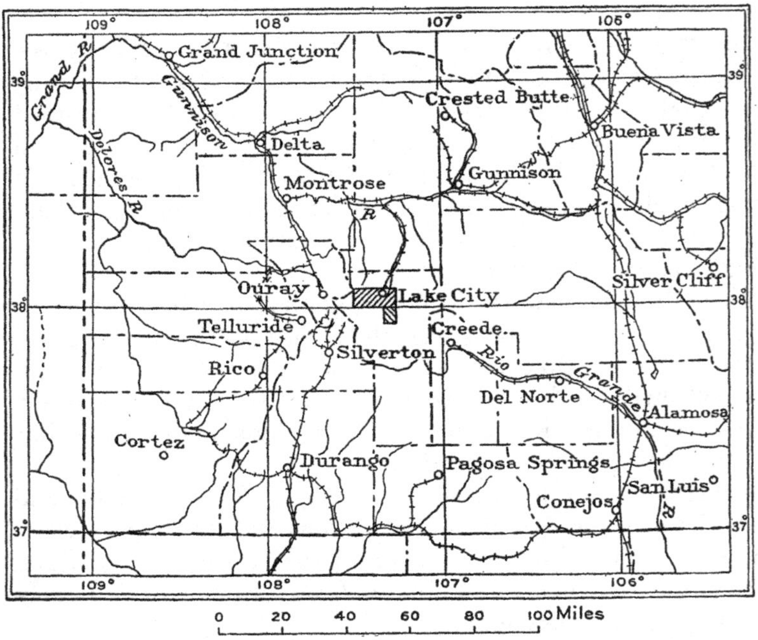

FIGURE 1.—Index map of a portion of Colorado, showing the location of the mining region discussed in this report.

mountains in Colorado, the most notable probably being Uncompahgre Peak, which stands 14,306 feet above sea level.

Lake City, a view of which is shown on Plate II (p. 12), is in the heart of the San Juan Mountains, in the northern part of Hinsdale County, at the junction of Lake Fork, a southern tributary of Gunnison River, and Henson Creek, which flows into Lake Fork from the west. The town is 8,663 feet above sea level, nearly the lowest point in the mining region here discussed. The ore deposits described are scattered along Henson Creek, to the west of Lake City, for about 10 miles and to the south, along Lake Fork and the western shores of Lake San Cristobal, for about 3 miles. The main region studied embraces an area of about 78 square miles, extending from latitude

38° to 38° 5′ N., and from longitude 107° 15′ to 107° 30′ W. To this is added about 3 square miles in the vicinity of Lake San Cristobal and about its shores. The location of these areas is shown in the index map forming figure 1 and in detail in Plates I and III and figure 20.

The region may be reached most conveniently through the valley of Lake Fork. Before the advent of the railroad a wagon road along Lake Fork served as the principal means of access, and the active development of the district from 1871 to 1889 was possible only under the stimulus of the discovery of rich ores. From other directions Lake City is accessible by wagon road from Creede and Silverton, and by a steep and rather hazardous trail from Ouray, on the west.

A branch of the Denver & Rio Grande Railroad from Lake Junction, 36 miles north of Lake City, now connects the town with the main narrow-gage line. From Lake City good wagon roads or trails lead to the principal mines; supplies are hauled as far as possible by teams and are then transferred to pack trains.

TOPOGRAPHY.

The surface of the region is composed principally of igneous rocks, which, under the action of streams, glaciers, and atmospheric agencies, have been wrought into a topography that is notable for its extreme roughness and irregularity. Viewing the country from Uncompahgre Peak, one is impressed by the approximate uniformity in the heights of the many high peaks of the San Juan Mountains and is led to conclude that these summits represent a formerly existing plateau which has been dissected by deep erosion.

The average difference in elevation between the stream beds and the ridges separating them is about 2,000 feet, but in some places the divide between two forks of the same small stream rises to heights of 3,000 feet or more above the bottom of the gulch. The lowest altitude in the quadrangle is slightly over 8,000 feet above sea level, and the elevation of Uncompahgre Peak, not over 12 miles distant, is 14,306 feet.

The region is in general drained toward the north, but the principal mining area is traversed by an eastward-flowing stream called Henson Creek, which, in the lower part of its course, runs through a deep, picturesque canyon. (See Pl. II, p. 12.)

OFFICIAL MINING DISTRICTS.

Hinsdale County is divided into mining districts, which here, as in other mining regions, have somewhat indefinite boundaries. As outlined by the legislative act of 1893, this part of Hinsdale County contains two districts, the Galena and the Lake. The Galena district, named from the presence of considerable quantities of the mineral of that name in the earlier exploited ores, extends westward along Henson Creek from Lake City to the Ouray and San Juan County lines. The Lake district, so called from its proximity to Lake San Cristobal, embraces the north and east portions of Hinsdale County. It extends southward from Lake City and westward along the valley of Lake Fork, including such prominent mines as the Golden Fleece, Black Crook, Isolde, and others in the section known as Burrows Park.

CLIMATE.

The effect of climate on the mining industry is fully appreciated by those who are engaged in mining operations in this region. In winter the streams are frozen and most of the mines are forced to use steam or gasoline engines instead of water power. The extra expense is great. Snow is so deep from December to April that operators of mines near timber line have to lay in supplies for the winter or cease work until spring.

TIMBER.

The timber in this region is all spruce or aspen, much of it available for use in the mines. The laws regarding the use and purchase of timber on mining claims, forest reserves, etc., are contained in brief in the "Use book" published by the Forest Service of the Department of Agriculture.

"Dry rot" of mine timbers is said to be a little more prevalent in the mines of the Lake City region than in most of the other mining districts in the vicinity. It is possibly due to the use of green timber in many of the workings; to insufficient circulation of air; to moisture in the shafts, adits, stopes, etc., sufficient to propagate fungous growth and not enough water to wash it away; and to the total neglect to use preservatives. The remedies for these evils are self-evident. If a few inexpensive precautions were taken, accidents to mine workings would greatly decrease. Circular No. 111 of the Forest Service gives a good brief account of methods of prolonging the life of mine timbers.

HISTORY.[1]

The history of the Lake City mining region is one of alternations—of general depression and of excessive activity—which have rendered its existence a little more eventful than that of the neighboring towns in the San Juan Mountains. These alternations have been due to several causes, but chiefly to the extreme richness of a few of the ore bodies discovered and the poverty of the rest. The periodical discoveries of new ore bodies of promising appearance were immediately followed by great inrushes of all sorts of people, whose presence

[1] Based mainly on notes furnished by Mr. J. J. Abbot, some of which are quoted directly.

A. VIEW OF LAKE CITY, COLO.

Photograph by Whitman Cross.

B. VIEW OF NORTH SIDE OF HENSON CREEK.

Photograph by Whitman Cross.

made the country thrive for a time. Similar variant conditions have prevailed to a greater or less extent in almost all mining centers, but in few places in Colorado have they been so pronounced as at Lake City.

Precious metal was probably first discovered in the Lake City area about 1842 by a member of the Fremont party, but no one, not even the explorer, has been able to locate the place or even the stream from which the first small amount of gold was panned. On August 27, 1871, with the discovery of the Ute and Ulay veins by Harry Henson, Jorl K. Mullin, Albert Meade, and Charles Godwin, the history of Lake City began. At that time all of the land which is now the "San Juan" belonged to the Indians. The reports of mineral wealth brought many prospectors into the region, and the red men became very much irritated at the frequent encroachments upon their domain. Finally, in 1874, to avert open hostilities, a treaty was drawn up and ratified by the Senate, whereby a strip of land 60 miles wide and 75 miles long was ceded to the United States Government by the Ute Indians.

In August, 1874, Hotchkiss (the leader of the expedition that built a wagon road from Saguache to Lake City) discovered the rich vein now known as the Golden Fleece and named it the "Hotchkiss." News of the strike spread rapidly and Lake City soon became a center of activity, the county seat being removed from San Juan to Lake City, where it has remained. During the same year reduction works were erected at Lake City and a third stamp mill was built in the Summit mining district.

Development was continued and new discoveries were made almost daily. The first boom attained its climax in 1876, coinciding with the opening up of Ocean Wave Group and the continued production of the Hotchkiss and the Ute and Ulay mines. During the spring the erection of a concentrator was begun and ground was broken for a smelter at the falls just above the city. Soon afterwards the reaction and "lull" so characteristic of the region began.

During the next three years work was continued on the Ute and Ulay and the Ocean Wave properties, the Excelsior mine was located (April, 1878), and the Crook and Ocean Wave smelters were completed.

The year 1880 marked the beginning of the biggest boom in the Lake City region. A great deal of work was done on the Palmetto group, which lies just west of the Lake City quadrangle. The St. Louis, Capitol, Czar, Silver Chord, Young America, Yellow Medicine, Pride of America, Vermont, Red Rover, and many other properties near Capitol City were being worked with various results.

Probably the most talked-of find during this period was that of the Golden Wonder, in Deadman Gulch, so named from four men who were killed and partly eaten there by their companion, Alfred Packer, in the winter of 1874.

This second boom period reached its climax near the close of 1881. In that year the Denver & Rio Grande Railroad, which had started to build to Lake City, became financially involved and ceased construction. In the fall of 1883 the Ute and Ulay shut down and for four years Lake City was practically dead.

In 1887 considerable ore was shipped from the Ulay, Vermont, and Yellow Medicine properties. The shipments from the Yellow Medicine mine fell off perceptibly in 1888, but the Ulay and Vermont continued to ship large quantities of ore. The Gallic was discovered during this period and later made a few shipments.

In 1889 the branch railroad was completed and soon afterward very rich ore was reported from the Golden Fleece. A single car of petzite ore from this mine is said to have yielded $50,000. The extreme richness of this ore stimulated mining throughout the region for about ten years. The total output of the Golden Fleece mine has been $1,400,000.

In 1890 some 20 mines in the Lake City quadrangle were shipping ore. During 1891 the Ute and Ulay alone produced over $400,000 and the total production from these mines has been $12,000,000.

In the late nineties the mining activity in the region almost reached a boom. Much work was done upon other properties in the vicinity of the Golden Fleece and some ore was shipped. The Golden Fleece Extension, Lake View, Black Crook, Contention, and others were operated. In June, 1897, the first ore was extracted by the present owners from the Hidden Treasure mine, as much as 22,000 tons of ore per year having been shipped from this property, which still continues to produce intermittently. The Czar first became a regular shipper in 1899, but its active life was short. The crest of the last wave of activity was reached during 1899, since which time operations have been more or less spasmodic.

PRODUCTION.

Hinsdale County originally included part of what is now Mineral County, in which the Creede mining district is located. On March 27, 1893, the legislature created Mineral County out of parts of Saguache, Rio Grande, and Hinsdale counties. For this reason statistics of production earlier than 1893 include also the production of Creede, which has been deducted in order to determine accurately the output of the present Hinsdale County, or practically the mining district here discussed. These statistics, except those for a few mines making confidential reports, have been compiled from the returns from individual mines as given in the reports of the Director of the Mint, in the reports of the State Bureau of Mines of Colorado, and, since 1904, in Mineral Resources of the United States, published by the United States Geological Survey.

The following table and figure 2 show the production of the county since 1884:

Metallic production of Hinsdale County from 1884 to 1906.

[Figures derived from United States Mint reports, 1884 to 1896, inclusive; from reports of State Bureau of Mines, Colorado, 1897 to 1903; from Mineral Resources of the United States, published by the United States Geological Survey, 1904 to 1908.]

Years.	Gold.	Silver.	Copper.	Lead.	Zinc.	Total value.
		Fine ounces.	*Pounds.*	*Pounds.*	*Pounds.*	
1884	$2,500	156,967	....	....	....	$180,317
1885	....	....	....	....	....	....
1886	2,060	18,586	....	30,435	....	23,743
1887	5,214	94.546	13,545	657,400	....	128,432
1888	2,667	110.433	1,815	1,495,614	....	172,586
1889	1,680	18.673	40,000	244,500	....	33,849
1890	3,577	61.023	40,950	546,920	....	96,112
1891	20,594	156.841	3,636	5,441,380	....	442,384
1892	13,529	418.422	20,182	6,[illegible]25,747	....	653,107
1893	88,279	340.774	(a)	(a)	....	354,754
1894	95,293	404,750	(a)	(a)	....	350,286
1895	274,421	466,836	....	3,676,733	....	689,577
1896	215,648	465,598	13,006	6,934,099	....	725,668
1897	168,171	243,437	8,085	3,550,058	....	501,822
1898	51,282	186,456	104,038	9,828,482	....	529,151
1899	38,343	155,902	49,676	10,572,353	....	612,561
1900	56.470	155,485	29,180	9,377,062	....	600,309
1901	76,148	152,122	12,532	7,588,675	....	496,792
1902	98,348	117,177	8,314	6,213,763	....	428,733
1903	16,515	33.139	11,263	459,462	106,000	60,910
1904	7,692	39.283	10,530	1,054,421	75,815	81,416
1905	11,991	54.419	84,485	767,681	2,085	94,244
1906	24,510	87.940	63,261	753,950	30,475	140,543
1907	7,520	50.109	99,410	1,204,628	23,034	125,678
1908	2,454	29.498	188,698	280,465	....	54,776
Total	1,284,906	4.047.416	802,606	78,903,828	237,409	7,577,750

a No figures for lead and copper available.

No production is given for 1885 in the mint report, so the district was probably idle in that year. The production prior to 1884 can not be definitely determined, but as the Ute and Ulay mine was in active operation and the Ocean Wave and others were productive, it is probable that accurate statistics would show a yield nearly, if not quite, equal to that given in the table. This is the more likely as the years 1876 to 1881 marked the greatest boom that Lake City has ever experienced.

From these figures the curve in figure 2 showing the production of copper, lead, gold, and silver has been constructed. This curve shows clearly that the culminating years of the later history of the district were 1895 and 1896, in the latter of which the yield was nearly three-quarters of a million dollars. It shows the general rise, culmination, and decline of the district, and brings out clearly the alternate periods of activity and depression. In 1889 the production was less than $35,000, but the advent of the railroad into Lake City that year caused a rapid increase, culminating in 1892. In 1893 the financial depression, coupled with a gradual decline in the value of silver caused a sharp fall, followed by a rapid rise until 1896. The decrease in 1897 and 1898 was probably due to local conditions. From 1898 the increasingly rapid decline, which culminated in 1903, represents the rapid exhaustion of the upper portions of the mines, whose richness was due to oxidation and secondary enrichment. After 1903 a slight revival indicates the attempts to handle low-grade material. The year 1908 shows the lowest output since 1885, and it is probable that unless the discovery of new veins should cause new activity in this region very little further production is to be expected.

MINES.

The following is a list of the mines which have been worked in Galena and Lake mining districts and which have either yielded some ore or have shown evidences of mineralization. The arrangement is geographical.

The production of some of these mines, as shown by Mint reports, statements of owners, the Geological Survey's reports entitled "Mineral resources of the United States," and other sources of information, is given after the names. These figures serve to

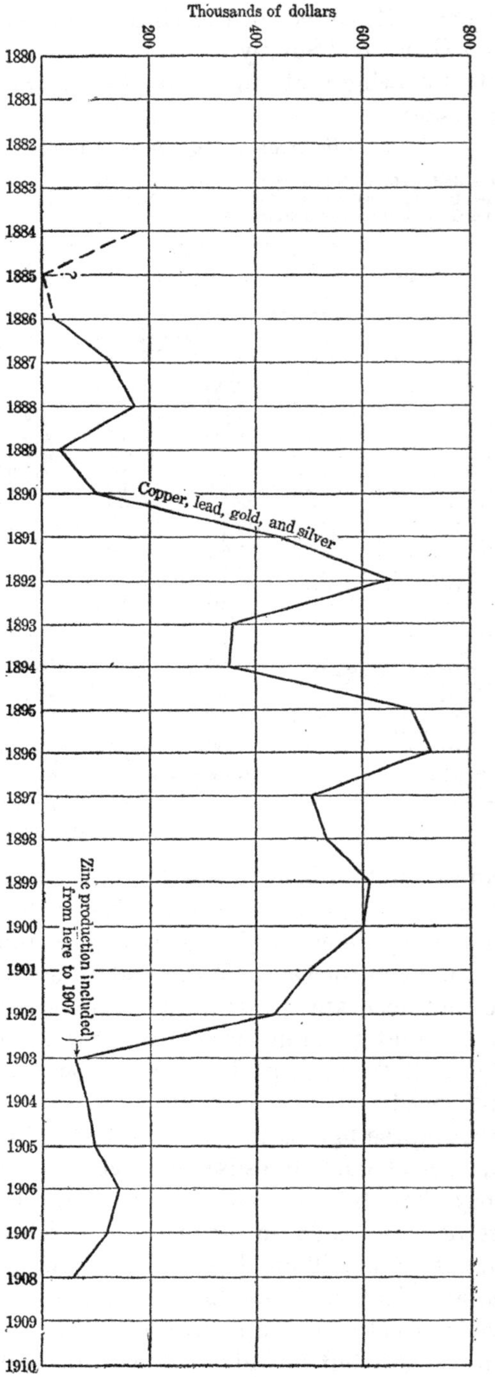

FIGURE 2.—Curve showing the metallic production of Hinsdale County, Colo., since 1884.

separate in a general way the important productive mines from the less productive and from mere prospects.

List of mines in Galena and Lake mining districts.

Capitol City.

Gallic and Vulcan.
Woodstock.
Ottawa.
Toby.
Lilly.
Capitol City ($27,990).
Yellow Medicine ($40,494).
Silver Chord Extension.
Silver Chord.
Czarina.
Czar.
Excelsior ($18,006).
Broker.
Lucky Strike.

Henson Creek (Main Stream.)

Vermont ($142,207).
Ocean Wave ($26,305).
Wave of the Ocean.
Lellie (Red Rover).
Baltimore.
Big Casino.
Little Casino ($420).
Pride of America.
Fanny Fern.
Cleveland.
Sacramento.
Yellow Jacket.
Ute and Ulay (estimated $12,000,000).
Hidden Treasure (estimated $700,000).
U. P.
Pelican.
Missouri Favorite.
California.

North Lake Fork (south of Lake City).

Nellie M.
Monte Queen.
Dauphin and Rob Roy.
Sulphuret and Cora ($3,090).
Golden Wonder.
Lode Star.
Ilma (Black Crook) ($124,447).
Belle of the West ($56,302).
Contention ($2, 332).
Golden Fleece (from mine records $1,400,000).
Carmi ($580).

Carson.

[These mines were not studied.]

Mayflower ($3, 590).
George III ($5,439).
Bonanza King ($3,290).
Hamilton.
St. Jacobs.
Maid of Carson ($27,397).

West Lake Fork.

[These mines were not studied.]

Isolde.
Tobasco.
Cleveland.
Black Wonder ($162).
Pelican ($2,084).
Bon Homme ($2,265).
Mount King.

South Fork of Henson Creek (south of Capitol City).

[These mines, except the Moro and Paymaster, were not studied.]

Wyoming ($775).
Independence ($4,520).
John Hough ($1,289).
Miners' Bank ($2,600).
Henson Creek ($700).
*Palmetto ($11,206).
Highland Chief.
*Bonanza.
*First National Bank.
*Frank Hough.
Anchor (probably a little).
Moro (considerable production).
Paymaster (no data).

Certain mines indicated in this list were not studied, as they lie too far outside of the district examined to warrant field work, and they are not further mentioned in this report. Many of them, however, have contributed to the production of Hinsdale County.

Those marked with an asterisk are within the northeastern portion of the Silverton district and have been described by Ransome.[1]

The observed conditions 300 to 400 feet below the surface lead the writers to believe that only the roots of veins are present in the district and that all ore bodies are likely to diminish greatly in value in depth. Hence the life of any mine is likely to be brief. The change in character of the ore and gangue minerals with depth in the Ute and Ulay, Hidden Treasure, Black Crook, and numerous others illustrates this point remarkably well.

GEOLOGY.

By WHITMAN CROSS.

The Lake City district is a small part of the volcanic San Juan region. Geologically it is particularly allied to the adjacent portion of the San Cristobal quadrangle on the south, which has been recently surveyed, and to the Silverton and Ouray quadrangles on the southwest and west, respectively, the geology of which has been described in published folios.[2] For this reason the geology of the San Juan region as a whole will first be briefly discussed, especially the important relations of the rock formations that occur in the Lake City area. Next the geology of the Lake City district itself will be described.

SAN JUAN MOUNTAINS.

GENERAL FEATURES.

The San Juan Mountains consist chiefly of surface volcanic rocks or of intrusive igneous masses, which now cover an irregular area of more than 3,000 square miles. This volcanic area extends from San Luis Park on the east to an irregular and abrupt western mountain front in the Telluride quadrangle. On the north the volcanic rocks reach out beyond the mountainous district proper, the lower lavas capping long low ridges between southerly tributaries of Gunnison River, in some of them extending to the edge of the Black Canyon. The southern border of the volcanic district runs from the Telluride quadrangle somewhat south of east and a broad arm crosses into New Mexico.

It is evident that the lavas once extended far beyond their present limits on all sides except the east. Between the San Juan and West

[1] Ransome, F. L., A report on the economic geology of the Silverton quadrangle, Colorado: Bull. U. S. Geol. Survey No. 182, 1901.

[2] Silverton folio (No. 120) and Ouray folio (No. 153), Geol. Atlas U. S.

Elk mountains, which lie north of Gunnison River, there was once a continuous covering of volcanic rocks, but these rocks did not necessarily come from a single central source. The Hayden geologic map of Colorado is approximately correct in its representation of the limits of this volcanic area, but it fails to give any idea of the complexity of the lavas, either as to character or age. The eruptions occurred during nearly the whole of Tertiary time, when a great many different kinds of lavas were poured out, building up a huge volcanic plateau. Eruptive activity was not continuous during the Tertiary period, but was broken by long intervals of quiet, during which extensive erosion materially changed the topography of the volcanic pile.

In the survey of the San Juan region, which now covers more than its western half, it has been found desirable to map and to describe the rocks of the principal eruptive periods in groups or series rather than to emphasize the occurrence of special rock varieties, and this method will be followed herein.

THE EARLIEST ERUPTIONS.

The commencement of volcanic activity was clearly later than the deposition of the Upper Cretaceous coal measures of southwestern Colorado, but somewhat earlier than any of the recognized Tertiary formations of the region. This is shown by the Animas formation, which overlies the coal measures at Durango and which consists largely of the pebbles and gravel of andesitic volcanic rocks. These beds carry fossil plants and scanty vertebrate remains which prove them to be of the same age as the Denver formation, at the base of the Front Range—that is, early Eocene or uppermost Cretaceous. The source of the volcanic materials of the Animas formation is not yet known.

SAN JUAN TUFF.

The Animas formation does not extend into the San Juan Mountains proper, and the lowest or earliest member of the great volcanic complex found in the western part of the region is a fragmental deposit which has been called the San Juan tuff. It is a more or less plainly stratified series of tuffs, breccias, or coarse agglomerates of andesitic rocks, in which no fossils have been discovered. No lava flows have been found interbedded with the fragmental deposits.

The San Juan tuff has a maximum observed thickness of 3,000 feet and forms notable deposits in the Ouray and Telluride quadrangles, though it appears in all other surveyed areas on the border of the volcanic district. It occurs in typical form in the Cimarron and Blue Creek valleys of the Lake City quadrangle.

The source of the andesitic rocks of the San Juan tuff was a mystery previous to the survey of the Lake City quadrangle. It now appears probable that a portion of the older volcanic mass from which those tuffs were derived occurs in the valley of Lake Fork a few miles below the mouth of Henson Creek. This mass has not yet been described and its extent is unknown. Its relations will be described in a forthcoming report on the Lake City quadrangle.

SILVERTON VOLCANIC SERIES.

The group of lavas succeeding the San Juan tuff are known as the Silverton volcanic series, as they have their greatest development in the Silverton quadrangle. These rocks are of special importance in the area treated of in this report. They consist of an alternation of andesites, latites, and rhyolites, in both flows and fragmental beds.

PICAYUNE VOLCANIC GROUP.

The earliest of these lavas, first observed in the Silverton quadrangle, is a dark augite andesite occurring in relatively small exposures in the Animas Valley and its minor tributary, Picayune Gulch, and named in the Silverton folio after the latter locality, the Picayune andesite. It is now known that the andesite of Picayune Gulch is but one of the rock varieties erupted in the first part of Silverton time and that the complex group of early lavas is most prominent in the country between Henson Creek and the head of the Lake Fork and in the valleys of these streams. The name Picayune will be retained for the group since it is already in use and since the relations of these rocks to the overlying rhyolite are well shown in the original locality.

The Picayune volcanic group consists largely of andesite of the type first observed, but includes also several kinds of more siliceous lava. The various elements are so irregularly associated as to make it clear that the exposed rocks of the group belong to a central volcanic mass made up of surface lava streams, pyroclastic breccia and tuff, and intrusive bodies. The massive rocks have in many places been shattered and dislocated by volcanic shocks. (See also p. 24.)

EUREKA RHYOLITE.

Among the more siliceous varieties of rock in the Picayune group is a rhyolitic lava which became more and more abundant until it was finally poured out in a thick flow or succession of flows which appear to have buried the older complex of Picayune rocks and now form a prominent and persistent element of the Silverton volcanic series. This rock is called the Eureka rhyolite, from the notable cliffs of it about Eureka, in the Animas Valley. It may generally be effectively described as a flow breccia, for it characteristically carries many small fragments of rhyolite and other rocks. Gray or pinkish exposures of it are continuous for several miles in the Animas

Valley, and appear very typically in the canyon of Henson Creek for 2 to 3 miles above Lake City.

BURNS LATITE.

The extensive flows of Eureka rhyolite were followed by an eruption of several varieties of latite lavas and tuffs more or less in alternation. The relations of these lavas are much simpler than those of the Picayune volcanic group and their order of succession is in most places plain. Two fine-grained tuffs, especially, are widely distributed in the Silverton quadrangle, one below and the other above the principal lava flows. This aggregation of flows and tuffs is called the Burns latite, from Burns Gulch, a branch of the Animas near Picayune Gulch. The rocks of the Burns occurring in the Lake City area are all tuffs.

PYROXENE ANDESITE.

The Burns latite is succeeded by dark fine-grained pyroxene andesite in a complex of flows and fragmental beds reaching a maximum thickness of nearly 3,000 feet in the central part of the Silverton quadrangle. These rocks are closely related to some of the Picayune lavas but have distinguishing characters. Massive flows of pyroxene andesite of this upper part of the Silverton volcanic series are present on the north side of Henson Creek near Capitol City, but they decrease in thickness eastward and disappear entirely west of Nellie Creek.

HENSON TUFF.

The uppermost member of the Silverton volcanic series is a greenish gray andesitic tuff, named Henson tuff, from its most notable occurrence on North Fork of Henson Creek, in the Ouray quadrangle. Its development in the Lake City area is very subordinate.

POTOSI VOLCANIC SERIES.

The long Silverton epoch of eruption was followed by a quiet time, during which erosion was very active; and then began another great succession of volcanic outbursts, producing lava flows and clastic deposits alternately. These lavas appear to have covered a larger area than the earlier ones. They are mainly latite and rhyolite, with some andesite, all of types different from the preceding ones, and preserved in much less altered condition. These rocks are called the Potosi volcanic series because of their typical exposure and notable thickness and variety in Potosi Peak, Silverton quadrangle.

The Potosi lavas are most widely distributed in the San Juan region. They occur very extensively in the central San Juan Mountains, San Cristobal quadrangle, and extend far down the northern slopes toward the Gunnison. They cap certain ridges north of Henson Creek and occupy much of the Lake City quadrangle farther north. Uncompahgre Peak presents a fine section of Potosi flows.

HINSDALE VOLCANIC SERIES.

The western San Juan region exhibits no lavas more recent than those of the Potosi volcanic series, but in portions of the San Cristobal and Uncompahgre quadrangles, and presumably in others not yet examined, a later succession of eruptions took place, producing a series of lavas differing notably from the products of earlier eruptions. It is plain that much erosion of the Potosi volcanic series occurred before the extrusion of these later magmas, which, so far as known, closed the long sequence of lavas in the San Juan region.

This newly recognized series of lavas ranges from a rhyolite very rich in quartz and alkali feldspar but poor in calcic feldspar and in all ferromagnesian minerals to a normal olivine plagioclase basalt. Between these two extremes are lavas of several types possessing some characteristics distinguishing them from earlier lavas.

No rocks of this recent series have been described in earlier folios or reports, and it is proposed to call them the Hinsdale volcanic series because of their important occurrences in Hinsdale County, of which Lake City is the county seat. The most extensive deposits of these lavas thus far discovered are on the divide between Lake Fork and Cebolla Creek, directly east of Lake City. The section exposed is nearly 1,200 feet thick. The lower 800 feet of the series occur within the area of the accompanying map on a spur from the divide. Other important localities for these rocks are Cannibal Plateau, northeast of Lake City; the Continental Divide, between Lake Fork and the head of Cebolla Creek; several summits between branches of Clear Creek; and the hill north of Lost Lake, the last three being in the San Cristobal quadrangle. All localities named are in Hinsdale County.

The Hinsdale is, like the Potosi and Silverton volcanic series, a set of lavas representing one of the major divisions of the San Juan volcanic history rather than a petrographic group. It is too early to sharply define its limits.

LAKE CITY DISTRICT.

GENERAL GEOLOGY.

Nearly all the rocks of the Lake City area (see Pl. III), belong to the Silverton volcanic series, the great intermediate member of the Tertiary volcanic complex. No earlier formation occurs here, and only subordinate representatives of the later Potosi and Hinsdale volcanic series occur in the district. Intrusive masses are numerous, but though these are obviously younger than the rocks which they penetrate their exact age relations are nowhere clear.

The canyon of Henson Creek, which is the most prominent topographic feature of the area, presents an excellent though by no means

A. ROUNDTOP MOUNTAIN, LAKE CITY MINING REGION.

Photograph by Whitman Cross.

B. POST-GLACIAL CANYON OF HENSON CREEK.

Showing gorge in Eureka rhyolite. Missouri Favorite Mine is in the center.

complete section of the prevailingly somber rocks of the Silverton series. On its southern side the lowest (oldest) member of the series, the Picayune group, predominates, and on its northern side several higher (younger) members are well shown. The general northerly dip of the lavas and tuffs affords proof that the center of eruption during the Silverton epoch of volcanic action was south of the Lake City area.

The restrictions of the map prevent representation of the fact that the various rocks of the Silverton series do not extend far northward beyond Henson Valley nor eastward beyond Lake Fork. This limitation is due to great erosion in the epoch preceding their eruption. The San Juan tuff and perhaps earlier massive volcanic rocks once existed where the Silverton lavas of Henson Creek now are. By their erosion a steep southward-facing slope or cliff was produced near the north boundary of the Lake City district. The Potosi rhyolites and latites cap the divide north of Henson Creek on either side of El Paso and Nellie Creek, resting on the uppermost rocks of the Silverton series; they are very prominent in the northern part of the Lake City quadrangle, forming plateau-like ridges with a gentle northerly dip. They are cut in the valleys of the east fork of the Cimarron, Blue Creek, Elk Creek, and other streams, exposing the underlying rocks; but these are not at all like the Silverton volcanic rocks of Henson Creek, consisting, instead, of a great volcanic breccia or semi-conglomerate of rude bedding, which extends west into the Ouray quadrangle, where it is shown to be the San Juan tuff, a formation entirely older than the Silverton series. These relations of the San Juan tuff and the Silverton rocks are not clearly exposed in the Lake City quadrangle, but they are exhibited in the Ouray quadrangle and have been described and illustrated in the folio dealing with that area. The Potosi lavas, therefore, overlap the boundary between the San Juan tuff and the Silverton series.

Although the canyon of Henson Creek cuts so deeply into the Silverton volcanic series, it does not actually penetrate below it at any point within the Lake City quadrangle. There is, however, evidence that granite and very old quartzite rocks lie probably at no great distance beneath the Picayune lavas on Henson Creek.

That granite is one of the foundation rocks is indicated by a small exposure of that rock 2 miles southwest of Capitol City in the bed of a southerly branch of Henson Creek, which it enters east of the Moro mill. This granite is on the north side of a fault of undetermined throw and the exposure is but a few yards in diameter. It is but 1½ miles from this point south to the large fault block of granite, the great part of which lies north of Whitecross, at the head of the Lake Fork.

Quartzite resembling the pre-Cambrian rocks of the Uncompahgre Canyon at Ouray occurs in a small exposure, surrounded by volcanic rocks at an elevation of 11,800 feet, a little more than a mile southeast of Capitol City. The exposure is insufficient to show whether this quartzite is a large block included in the volcanics or a pinnacle of an underlying rugged quartzite topography, but it does make it clear that rocks other than granite go to make up the prevolcanic complex of this vicinity. At several places in the San Cristobal quadrangle chaotic breccia masses in the Picayune volcanic group contain fragments of granite and quartzite, some of which are several yards in diameter.

SILVERTON VOLCANIC SERIES.

PICAYUNE VOLCANIC GROUP.

Occurrence.—The rocks of the Picayune volcanic group, together with intrusive masses, occupy the greater part of the area between Henson Creek and the Lake Fork to the south. They occur connectedly in the Henson Valley from near Capitol City to a point below the mouth of Alpine Gulch and in separated areas in Wade Gulch, on the slope between Wade and Alpine gulches, and on the west side of the Lake Fork opposite Deadman Gulch. The rocks rise higher on the south side of Henson Creek than on the north, owing to a general northerly dip of their upper surface, on which the Eureka rhyolite rests.

The rocks of the group are prevailingly dark in tone and cause somber cliffs in which a rude stratified arrangement of successive flows or beds can be recognized in many places. Among the best exposures for a study of the Picayune rocks are the steep cliffs at several points between El Paso and Modoc gulches, and in Wade Gulch near the quadrangle line. There is much landslide and forest cover on the south side of Henson Creek, although the rocks are well exposed in many localities.

General character.—The lavas and fragmental rocks here called the Picayune volcanic group once formed a great volcano, the center of which was situated a few miles south of Henson Creek, in the area between that stream and the upper part of Lake Fork. The rocks of the group are principally dark andesites, dacites, and quartz latites, with a variable amount of light-gray rhyolite. The petrographic distinctness and the sharp local boundaries of certain masses made it at first appear practicable and desirable to distinguish them on the map, and much time has been spent trying to do this. But with greater knowledge of this old volcano it has become clear that a petrographic division of the group can be mapped only at some future time, after long and painstaking study and with a topographic base of much detail and great accuracy. The lavas, tuffs, and breccias of the group are therefore mapped collectively under one color, and only the more important and definite intrusive rocks are distinguished.

The structure of the Picayune mass is most irregular. A large part of it is breccia—that is, it is made up of angular rock fragments. Some breccias are composed almost exclusively of one kind of rock and appear to have been made by the shattering of massive rock in place. Other breccias are simply chaotic accumulations of several different rock types, in large and small fragments. Well-bedded deposits of more or less worn, transported material also appear but are not continuous enough to express a prevalent structure.

Lava flows alternate with breccias or tuffs in many local sections, but these, too, possess little lateral extent. Intrusive masses of various shapes and sizes, consisting of rock varieties for the most part identical with some of the flows or breccia fragments, are, by reason of the prevalent irregularity of all masses, distinguishable only with the greatest difficulty and in favorable localities.

No regular succession of magmas of different character has been made out. Probably the oldest lavas of the Picayune center were andesitic and the later ones chiefly rhyolitic or latitic. Such appears to be the general rule, but apparent exceptions are numerous.

The Picayune rocks are more extensively altered than the younger lavas. This is quite natural in view of the fragmental character of much of the mass and the local extent of even the largest massive bodies. These conditions must have been favorable to the free circulation of solvent and mineralizing waters.

Typical exposures.—The development of the Picayune volcanic group in the Henson Valley may be best described by giving details of typical localities. One of these is the ridge between Nellie and Pole creeks, on the southeast face of which the Picayune rocks are exposed for about 1,500 feet. above the mouth of Nellie Creek. A well-stratified conglomerate and breccia, forming ledge exposures by the roadside at several places between Pole and Modoc creeks, is the base of the exposed section. This stratified rock appears at first to have nothing to do with the darker volcanic complex about it, but it is, in fact, only a phase of the rhyolitic tuffs or conglomerates found in local development in many places, at various horizons. The conglomerate is overlain by a coarsely stratified breccia, dipping 25° to 30° N. Adjacent outcrops are similar for the most part but differ in relative proportions of rock types and in texture. At about 200 feet above Henson Creek a brecciated nearly homogeneous rock is overlain by a flow, the lower zone of which is marked by fluidal structure. This extends for about 100 feet vertically, the dense dark rock being almost homogeneous though more or less brecciated. At an elevation of 300 feet a finely bedded sandy tuff occurs with a flow only 5 to 6 feet thick in its midst. This flow has sharp contacts and fluidal border zones. A variable dip reaching 30° N. as a maximum is here distinct. Above these gray tuffs comes another much thicker flow similar to the one in the tuffs. Breccia and irregular masses of more basic rocks follow, with gray latite or rhyolite at certain horizons, and this alternation continues up to the massive Eureka rhyolite flow.

The Picayune rocks east and north of Capitol City deserve particular mention, for their relations to the Eureka rhyolite are not fully understood. The heavy coating of glacial débris on North Fork of Henson Creek obscures the geology very much, but the Lucky Strike tunnel, at Capitol City, penetrates greenish quartz-bearing andesite, quartz monzonite porphyry, and rhyolite similar to the Eureka type, the whole indicating a complex phase of the Picayune group.

The same rocks and other types known in the Picayune are found in Yellowstone Gulch and are encountered in the Gallic tunnel. In the latter locality Eureka rhyolite occurs on the slope above the tunnel and is the only rock observed along the banks of North Fork of Henson Creek, as far as Capitol City. These facts show that an arm of the Picayune group rocks extends up North Fork, rising much higher on the northeast side.

The rock referred to as quartz monzonite porphyry is no doubt an intrusive, identical in composition with the body represented on the map as penetrating Eureka rhyolite on the slope west of Capitol City. It is not improbable that the coarse quartz monzonite porphyry of the Gallic tunnel, Yellowstone Gulch, the Excelsior mine, and the Lucky Strike workings, as well as of certain outcrops surrounded by glacial material southeast of Capitol City, is of the same age as the body in Eureka rhyolite. But the boundaries can not be determined and so it can not be separated from the Picayune.

East of Capitol City and north of Henson Creek the line between Eureka rhyolite and the Picayune volcanic group has not been accurately determined and hence no definite boundary is shown on the map.

Some of the later rhyolite flows of the Picayune epoch were very much like the succeeding Eureka rhyolite and they do not seem to be so much broken as earlier lavas, so that it is difficult to determine whether the rhyolite of some areas south of Henson Creek should be referred to the Eureka or not. A massive gray rhyolite on the ridge west of Alpine Gulch closely resembles the Eureka, but it dips beneath dark andesitic rocks near the mouth of the gulch and is for this reason included in the Picayune.

EUREKA RHYOLITE.

The Eureka rhyolite in a succession of flows covers the Picayune volcanic group on the north side of Henson Creek and extends up the south slopes east of Alpine Gulch and near Capitol City. The character of the rock is well shown in the canyon of Henson Creek

(Pl. IV, *B*) for 2½ miles above Lake City and in the cliffs west of the town. It is also present east of the town on the lower slopes.

The Eureka rhyolite is typically a dull ash-gray or green felsitic rock with few phenocrysts that exhibits a pronounced fluidal texture as seen in exposures a few feet in diameter if not in hand specimens. Feldspar and biotite crystals give the rock a porphyritic facies, as a rule, the groundmass strongly predominating in most places. Inclusions of rock fragments very similar to the Eureka are common in most flows and suggest the term flow breccia. These fragments are usually but a fraction of an inch in diameter and they are commonly rather flat flakes arranged in bands, emphasizing the fluidal structure.

Some of the Eureka rhyolite flows are several hundred feet in thickness and their lateral extent is measured by miles. A small amount of tuff separates some of the flows, but more commonly these are in direct contact and are so similar that the boundaries are often not easily detected.

As has been pointed out (p. 20), the Eureka lavas are practically identical in character with some of those in the Picayune group. The justification for distinguishing the Eureka as a map unit lies in its mass and areal importance. The lava floods of Eureka time covered a large part if not all of the complex Picayune volcanic group.

BURNS LATITE TUFF.

The Burns latite tuff division of the Silverton volcanic series is represented in the valley of Henson Creek only by fine-grained sandy tuffs and calcareous beds, which extend northeast from the south base of Sunshine Mountain to the ridge west of Nellie Creek and beyond the area mapped. Near Nellie Creek the tuffs rest on Eureka rhyolite, in their normal relation, but in El Paso Gulch, Empire Mountain, and North Fork of Henson Creek they are separated from the rhyolite by intrusive sheets of pyroxene andesite. The tuff band varies greatly in thickness, owing to erosion before it was covered by andesite flows; northwest of Yellowstone Gulch it wedges out entirely.

In the drainage areas of Modoc, Findley, and Crystal creeks the Burns latite tuff is concealed by flows of the Potosi volcanic series, but it reappears west of Crystal Peak and occupies a considerable area between Slaughterhouse Gulch and Larson Creek. It is, however, much obscured by landslide material consisting largely of gray-green Burns tuff, which extends to Lake Fork immediately below Lake City.

The tuff reappears east of Lake City, lying on the Eureka rhyolite, and continues southward across Deadman Gulch, being interrupted by glacial débris and penetrated by several intrusive rocks.

The Burns tuffs are mainly fine-grained gray or greenish sandstones, made up chiefly of particles of volcanic rocks. Some beds include rock fragments several inches in diameter. Other layers are thin laminated shale. Certain layers of some exposures are rich in carbonate of lime, and true limestone strata 2 or 3 inches thick are locally developed. These calcareous beds are well exposed on the flat summit of the ridge east of El Paso Gulch. A few fossil leaves have been found in the Burns tuff in the ravine north of North Fork of Henson Creek, and fragments of carbonized plant stems are common.

PYROXENE ANDESITE.

Dark massive flows of pyroxene andesite normally succeed the Burns tuff in the Silverton volcanic series. Such rocks reach a thickness of several thousand feet in the Silverton quadrangle, but they do not exceed 800 feet in the valley of North Fork of Henson Creek and gradually thin out eastward. They have not been found east of Nellie Creek. These andesites occur in flows varying from a few feet to more than 100 feet in thickness.

Below the Burns tuff occur nearly identical sheets of rock, which have not been distinguished on the map. In several places it is clear that these lower bodies have been intruded, and they no doubt belong to the same epoch of eruption as the flows above the tuffs. In El Paso Gulch and the ridge east of it the evidence of intrusion is plain and is expressed in part on the geologic map.

The pyroxene andesites are dark porphyritic rocks, with numerous crystals of augite and plagioclase visible to the unaided eye. They are massive except in the upper zones of most flows, where a pronounced vesicular texture is commonly developed. Secondary quartz and bluish chalcedony are usually found in these vesicles and, as they readily weather out of the rock in exposed places, small nodules or fragments of them strew the ground in many places near outcrops.

The microscope shows that hypersthene was a former constituent of these rocks, its place being now taken by serpentine, chlorite, and other alteration products. Quartz and orthoclase are generally present in small amount.

HENSON TUFF.

The last member of the Silverton volcanic series is a tuff much like the Burns latite tuff. Its greatest development is in the valley of North Fork of Henson Creek in the Ouray quadrangle. It lies on the pyroxene andesites in the region west of Nellie Creek, but is not found east of that stream.

The Henson tuff is made up chiefly of débris of pyroxene andesite, but contains particles of other volcanic rocks; locally it contains angular fragments of rhyolite, latite, and andesite. There are no calcareous layers in it, like those of the Burns tuff, and no fossil leaves have been discovered.

POTOSI VOLCANIC SERIES.

The lavas of the Potosi volcanic series appear in the Lake City area only on the high ridges on either side of Henson Creek and east of the Lake Fork. They lie on various older formations, in some places with unconformity, testifying to much erosion following the deposition of the Henson tuff.

North of Yellowstone Gulch the Potosi lavas rest on Henson tuff or an intrusive quartz latite sheet, and in Empire Mountain they lie on pyroxene andesite and Burns latite tuff. On either side of Crystal Creek the Eureka rhyolite is the underlying formation, and east of Alpine Gulch the Potosi lavas come in contact with the Picayune rocks.

The Potosi lavas of the region north of Henson Creek are light or dark gray quartz latites exhibiting many soda-rich plagioclase feldspar phenocrysts and some of sanidine in a groundmass rich in quartz and sanidine. Biotite and augite are original constituents, but are generally quite decomposed.

The alteration of Potosi rocks by which they are silicified or kaolinized is extreme in the so-called "Iron beds" northwest of Broken Hill. Much of the rock is stained a brilliant red or yellow through oxidation of the pyrite which impregnated it at the time of its decomposition. Alteration of this type is also exhibited in less degree south of Broken Hill.

The Potosi rocks east of Alpine and Wade gulches are quartz latite porphyry belonging to a large body which apparently filled a great hollow. This rock is characterized by more prominent crystals both of plagioclase and sanidine feldspar than are found in the common thinner flows north of Henson Creek. The upper parts of Crown Mountain and Red Mountain in the San Cristobal quadrangle are made up of this same phase of the Potosi quartz latite.

The Potosi lavas east of Lake City are mainly dark pyroxene andesite of a type common to the southeast. North of Horse Park they are representatives of quartz latite flows beneath the andesites.

HINSDALE VOLCANIC SERIES.

The lavas of the Hinsdale volcanic series overlie the Potosi flows east of Horse Park. They belong to the succession of rhyolite, basalt, and intermediate rocks forming the upper portions of the divide between Lake Fork and Cebolla Creek. The basalts are the capping rocks of Cannibal Plateau and are recognized as such in the Hayden reports and on the map. These rocks have not as yet been thoroughly examined, and as they are supposed to be more recent than the ore deposits of the Lake City region they need no further discussion in this place.

INTRUSIVE ROCKS.

The intrusive rocks of the region are distinguished on the map under three colors. These rocks are not intimately related to the ore deposits and will be passed over in this place with but brief mention. Some of the types occur in more important masses in the areas north or south of that with which this report deals and will be described in full in forthcoming reports. The various kinds of intrusives may be conveniently grouped under the heads of the map legend.

Rhyolite.—The most widely distributed type in this group is a rhyolite which is very abundant south of Henson Creek in bodies of various sizes and shapes, cutting the Picayune volcanic group. To the south, as far as the Lake Fork, in the San Cristobal quadrangle, this rhyolite is very abundant throughout the area of Picayune rocks and makes up a large part of some of the highest summits, such as Red Cloud and Sunshine peaks. It includes many large blocks of the Picayune lavas, a few of which are represented on the accompanying map.

This rhyolite is a grayish porphyry exhibiting phenocrysts or orthoclase, and quartz with a few biotite flakes, in a felsitic groundmass. In some places the rock has a strongly marked fluidal texture and in others it is massive. Inclusions similar to those in the Eureka rhyolite are abundant in some localities. The groundmass is very fine grained in places but is never glassy. A gradation to coarse-grained texture takes place in certain masses, so that some of the rock may well be called granite porphyry. Masses of such texture occur principally in Alpine Gulch, in the San Cristobal quadrangle.

This rhyolite differs markedly from the Eureka rhyolite in the constant presence of quartz phenocrysts penetrated by many white arms or embayments of the groundmass. It is also more typically a rhyolite than is the Eureka, being generally free or nearly so from crystals of lime-soda feldspar.

A rhyolite porphyry characterized by numerous phenocrysts of smoky quartz and clear sanidine, but almost destitute of any ferromagnesian mineral, occurs in many crosscutting dikes and irregular bodies in the drainage areas of Nellie and El Paso creeks and the North Fork of Henson Creek. Several small masses of this rock are represented on the map. This rhyolite penetrates flows of the Potosi volcanic series and is probably much younger than the similar rock in the Picayune area, already referred to.

A rhyolite of felsitic type forms the summit of the hill east of Lake City, which stands at an elevation of 10,726 feet. It is fine-grained, gray, or pinkish in color, and has a fluidal texture due to the arrangement of spherulitic bands. A smaller mass occurs in the knoll north of Deadman Gulch and seems to be intrusive, though surrounded largely by glacial gravels.

The intrusive mass west of Crystal Creek opposite Sugarloaf Rock is a rhyolite resembling the Eureka rhyolite in many particulars.

Quartz latite.—In the hill east of Lake City and in Deadman Gulch are several sheets of a fine-grained quartz latite, which cut the Burns latite tuff irregularly. This rock is gray in color, with small crystals of plagioclase, sanidine, biotite, and quartz in a subordinate groundmass. The quartz crystals are embayed by tongues of the groundmass, after the fashion of the rhyolite south of Henson Creek, but the rock contains much plagioclase, is rich in biotite, and carries a little hornblende. It is called quartz latite porphyry.

Another rock of this kind, nearly identical in character with certain flows of the Potosi volcanic series, occurs as a sheet or sill injected above, below, or within the Henson tuff, in the area between North Fork of Henson Creek and Nellie Creek. In some places it is 200 to 300 feet thick. It extends north 2 miles from Broken Hill, reappearing at the head of Cimarron Creek, southwest of Uncompahgre Peak. This rock is gray, with prominent biotite crystals, and is more compact than the usual Potosi flows, but it shows fluidal texture in some places and may represent an intrusion during the Potosi epoch of a magma which reached the surface elsewhere in the vicinity.

The quartz latite forming the summit of Sugarloaf Rock is a fine-grained porphyry different from any of the preceding types. It contains both hornblende and biotite phenocrysts, while quartz is restricted to the groundmass.

Andesite.—A sheet of dark fine-grained andesite cuts obliquely across the Burns latite tuff in the face of the hill east of Lake City. The rock carries hornblende in abundance, with some augite and biotite, all greatly altered. No other mass of this character occurs in the area described in this bulletin.

Quartz monzonite porphyry.—In the ridge west of Capitol City there occurs a branching intrusive body of much more coarsely crystalline texture than is exhibited by any other intrusive distinguished on the map. It cuts the Eureka rhyolite and is probably more extensive than is indicated by the map, for landslide and glacial débris obscure its outcrops very greatly.

This rock contains many prominent hexagonal tablets of biotite, associated with plagioclase and some quartz phenocrysts. The groundmass is rich in orthoclase and quartz of microgranular texture. Rock of nearly identical character occurs in the breast of the Gallic tunnel, according to Bancroft, and in all probability this mass is connected beneath the surface with the body mapped.

Decomposed porphyry of Deadman Gulch.—The highly decomposed rock occurring in Deadman Gulch and extending into the adjacent valley on the south is in part a porphyry with little remaining of its original constituents except quartz phenocrysts with penetrating arms of the groundmass. This suggests that there may be here intrusive bodies of quartz latite or rhyolite porphyry similar to some one of the types described, but in a large part of the area the rock is so completely silicified or kaolinized that the primary characters have wholly disappeared. Since to these obscuring effects of decomposition are added the covering of all but the upper contacts by glacial or landslide detritus, it has been impossible to determine the original character of the rock or even to decide whether or not more than one rock type is represented.

This area of alteration is directly connected with that at the head of the great Slumgullion mud flow which dammed the Lake Fork and caused Lake San Cristobal. The extreme head of that flow is very near the southeast corner of the Lake City quadrangle. The alteration in Deadman Gulch is not quite so thorough as in the Slumgullion Basin, but disintegration of the decomposed rock is a common feature of prospect dumps in the former area.

The decomposition in question extends upward with diminished intensity into the lavas of the Potosi volcanic series, but apparently the massive flows of andesite served to confine the decomposing waters in some degree and thus promote the alteration of underlying rocks.

ORE DEPOSITS.

SAN JUAN REGION.

GEOGRAPHIC CONTINUITY.

The Lake City area is not an isolated locality whose general relations may be discussed without reference to those of adjacent regions. On the west is the Ouray region; on the southwest the famous and productive Silverton and Telluride regions; and still farther southwest are the well-known camp of Rico and the lesser districts of the La Plata and Needle Mountains quadrangles. These six mining regions together make up the major part of the famous San Juan mining region and form one connected mountainous district characterized throughout by extremely rugged topography, a preponderance of extrusive and intrusive eruptive rocks, and an extensive and more or less closely related mineralization. The Lake City district is situated on the northern border of this region and is separated from the Creede district to the southeast by a considerable tract of country in which no ore bodies of consequence have yet been discovered. The veins differ from those of the Telluride, Ouray, and much of the

Silverton region in being in an older series of volcanic rocks; the San Juan tuff, which is so abundantly mineralized in the latter quadrangles, does not outcrop within the mineralized part of the Lake City area. The region is geologically continuous, however, with the eastern and central portions of the Silverton quadrangle, a great part of the rocks there exposed belonging to the same groups as those in the vicinity of Lake City. The veins show many close similarities, both physically and mineralogically, to those in the adjoining portion of the Silverton area; and it is highly probable that they owe their origin to similar conditions of mineralization, fissure formation, etc.[1]

GENESIS OF MINERALS IN THE SAN JUAN REGION.

It has for some time been recognized that different associations of minerals form under different conditions of temperature and pressure. Those formed under high temperature and pressure are in extreme instances entirely different from those formed under low temperature and pressure, and the mineral formation may be regarded as a measure of the vertical depth at which ore formation has occurred; that is to say, the earth's crust in any given locality may be divided into zones of depth, and within the vertical range of each zone characteristic temperature and pressure may be assumed to have existed. The mineral contents of the veins of any region will then express in some degree the depth at which the minerals formed. In 1907 Lindgren[2] pointed out the existence of certain zones and set forth the minerals characteristic of each of them. In 1908 W. H. Emmons[3] followed with a paper giving a tentative genetic classification of minerals, amplifying the work of Lindgren.

In comparing the different districts of the San Juan Mountains with the Lake City district the writers have endeavored, by the use of this work of Lindgren and Emmons together with their own additions and observations, to classify the minerals constituting the ore deposits of the five districts above mentioned into groups, each of which is characteristic of a particular zone.[4] Six such groups were found to exist, as follows: (1) Minerals of the oxidized zone, (2) minerals due to secondary sulphide enrichment, (3) minerals formed at moderate and shallow depth, (4) contact metamorphic minerals, (5) minerals of the deep-vein zone, (6) minerals which may occur in all or most zones and are of no diagnostic value.

---

[1] Ransome, F. L., A report on the economic geology of the Silverton quadrangle, Colorado: Bull. U. S. Geol. Survey No. 182, 1901.

[2] Lindgren, W. L., The relation of ore-deposition to physical conditions: Congr. geol. intern., Compt. rend. 10e sess., Mexico, 1906, pp. 701-724, 1907; Econ. Geology, vol. 2, No. 2, pp. 105-127, 1907.

[3] Emmons, W. H., A genetic classification of minerals: Econ. Geology, vol. 3, No. 7, pp. 611-627, 1908.

[4] See tables, pp. 34 and 46, compiled to show the comparative mineralogy of the San Juan and the primary mineralogy of the Lake City region.

The first two groups belong to surficial zones and afford no true basis of genetic comparison, as they may be superposed on any series of deposits, of whatever origin, which contain sulphides and other elements in necessary amount. Differences in them afford comparison only of varying topographic and climatic conditions. As such conditions vary but little in the San Juan region, the secondary and oxidation products show a striking similarity throughout. The third, fourth, and fifth groups furnish a good basis for a comparison of the primary minerals. The details of their occurrence (see table below) in the various districts of the San Juan region, though accurate only within the limits of our present knowledge, yield some significant results, which in the writers' opinion, serve to emphasize the close relation of the Lake City region and the other areas and to bring out such differences as exist.

TABLE 1.—*Comparative mineralogy of the San Juan region.*

	Lake City.	Ouray.	Silverton.	Telluride.	Rico.
1. Minerals of oxidized zone.		Basic ferric sulphate.			
	Kaolinite	Kaolinite	Kaolinite	Kaolinite	
	Limonite	Limonite	Limonite	Limonite	
	Hematite	Hematite	Hematite	Hematite	Oxidized products not specifically enumerated.
	Native silver	Native silver	Native silver		
	Gold	Gold	Gold (?)	Gold	
	Copper		Copper	Native copper	
	Malachite	Malachite	Malachite	Malachite	
	Azurite	Azurite	Azurite	Azurite	
	Cerussite	Cerussite	Cerussite	Cerussite	
	Anglesite		Anglesite	Anglesite	
	Chalcanthite			Chalcanthite	
	Pyrolusite	Gypsum		Gypsum	
2. Minerals produced by secondary sulphide enrichment.	Covellite				
	Chalcocite	Chalcocite	Chalcocite		
	Bornite		Bornite		
	Galena				
	Proustite	Proustite	Proustite	Proustite	Proustite.
	Pyrargyrite	Pyrargyrite		Pyrargyrite	
	Argentite		Argentite		Argentite.
			Stromeyerite	Stephanite	Stephanite.
				Polybasite	Polybasite.
3. Minerals of moderate and shallow depths.	Sericite	Sericite	Sericite	Sericite	
	Hinsdalite		Hinsdalite		
	Jasperoid	Jasperoid	Jasperoid	Jasperoid	Jasperoid.
	Barite	Barite	Barite	Barite	Barite (rare).
	Rhodochrosite	Rhodochrosite	Rhodochrosite	Rhodochrosite	Rhodochrosite.
		Rhodonite	Rhodonite		
	Tetrahedrite	Tetrahedrite	Tetrahedrite	Tetrahedrite	Tetrahedrite.
			Enargite		
			Bournonite		
			Zinkenite		
			Guitermanite		
			Hübnerite		
	(Bismuth compounds).		Bismuthinite		
4. Contact metamorphic minerals.		Brown garnet			Garnet.
		Epidote			Epidote.
		Actinolite			
		Tremolite			
		Magnetite			Magnetite.
					Vesuvianite.
					Wollastonite
					Specularite.
					Chlorite.
Minerals of the deep vein zone.			Specularite		
				Zoisite	
				Spinel	
				Picotite	
				Magnetite	
				Biotite	
				Garnet	

TABLE 1.—*Comparative mineralogy of the San Juan region*—Continued.

	Lake City.	Ouray.	Silverton.	Telluride.	Rico.
6. Persistent minerals				Apatite	
	Quartz	Quartz	Quartz	Quartz	Quartz.
	Calcite	Calcite	Calcite	Calcite	Calcite.
		Dolomite	Dolomite	Dolomite	
	Fluorite		Fluorite	Fluorite	Fluorite.
	Chalcopyrite	Chalcopyrite	Chalcopyrite	Chalcopyrite	Chalcopyrite.
	Galena	Galena	Galena	Galena	
	Sphalerite	Sphalerite	Sphalerite	Sphalerite	Sphalerite.
	Stibnite		Stibnite		
			Molybdenite		
	Tellurides	Tellurides	Tellurides		
		Gold	Gold	Gold	
				Arsenopyrite	
	Pyrite	Pyrite	Pyrite	Pyrite	Pyrite.
				Siderite	

In placing the rich silver minerals proustite, pyrargyrite, stephanite, and polybasite in the list of those produced by secondary sulphide enrichment the writers have followed mostly the results of personal observations, but to some extent also the descriptions of Ransome [1] and Purington.[2] Proustite and pyrargyrite are without question of secondary origin in the ore deposits of Lake City and Ouray (pp. 62–63). From Ransome's and Purington's descriptions it seems probable that they are secondary in Rico and in Silverton also. Stephanite and polybasite are described by Purington and Ransome as the last-formed minerals of the veins in which they occur, and in the writers' judgment they should also be placed in the secondary sulphide column for both the Silverton and Telluride districts.

MINERALOGICAL SIMILARITY.

Disregarding for the moment the contact metamorphic deposits, a general view of the table shows that the veins of the San Juan region exhibit a fairly close mineralogical similarity in their most common primary constituents. They are characterized chiefly by pyrite, argentiferous galena, sphalerite, and tetrahedrite, with a gangue composed largely of quartz with subordinate rhodochrosite and other carbonates. Little of the argentiferous galena is rich in silver in any of these districts unless it is accompanied by tetrahedrite or some rich secondary silver mineral. Gold is commonly subordinate in value to silver, notable exceptions being in the Golden Fleece, Camp Bird, and some other mines. Fluorite is rare and almost lacking in the Lake City region. Chalcopyrite is almost universally present in some part of every vein, but generally in less amount than the other minerals. The relative abundance of these primary minerals, of course, varies locally and could serve as a basis

[1] Ransome, F. L., A report on the economic geology of the Silverton quadrangle, Colorado: Bull. U. S. Geol. Survey No. 182, 1901; Ore deposits of the Rico Mountains, Colorado: Twenty-second Ann. Rept. U. S. Geol. Survey, pt. 2, 1901, pp. 229 et. seq.

[2] Purington, C. W., Preliminary report on the mining industries of the Telluride quadrangle, Colorado: Eighteenth Ann. Rept. U. S. Geol. Survey, pt. 3, 1898, pp. 751 et seq.

for partial subdivision, but the differences are either in the minor constituents which characterize those minerals now known to be secondary (stromeyerite, argentite, etc.), or represent merely local preponderances.

In the Lake City and Ouray regions (columns 1 and 2 of Table 1) all minerals characteristic of the deeper vein zones are completely lacking. In the Rico region, except in certain contact-metamorphic deposits, the veins again show a striking want of any minerals characteristic of deep-vein formation. In the Silverton region specularite is a not uncommon vein mineral, perhaps indicating an approach to a deeper zone of mineralization than is characteristic of the Lake City lodes. In the Telluride region the presence of zoisite, spinel, picotite, magnetite, biotite, and garnet, associated with the minerals which characterize the veins throughout the region, seems to indicate formation at greater depths than in the other districts. This is not absolutely certain, however, as the exact nature of the deposits of these minerals is not stated by Purington.

In both the Rico and Ouray regions contact-metamorphic deposits occur where intrusive monzonite intersects or is in close proximity to limestone. These deposits are characterized by magnetite, pyrite, sphalerite, chalcopyrite, wollastonite, garnet, epidote, specularite, and chlorite. Deposits of this type have with little doubt originated under conditions of high temperature and great pressures, and they are usually regarded as belonging to the deep zone. There seems no good reason to believe that in the Ouray region they represent a period of ore deposition very widely separated from that of the normal lodes and replacements, but they were probably formed long enough before the overlying cover of rock had been sufficiently eroded and the upper part of the monzonite mass become sufficiently cooled to permit the formation of the veins of shallower depth that chiefly characterize these districts and in many places intersect the monzonite itself. The added temperature and pressure which these contact-metamorphic deposits imply would then have been caused by the aggressive entrance of the heated intrusions of monzonite. The first effect would be the formation of contact deposits where this rock had intersected limestones. The subsequent cooling of the more deeply buried mass would have permitted the escape of the vein-forming vapors which produced the normal lodes. For these reasons the contact-metamorphic occurrences cited seem to the writers to have little bearing on the depth of the general ore formation of the region, but to have been rather the result of local conditions.

In general, then, the Lake City lodes may be regarded mineralogically as the outer or northeasterly edge of the heavily mineralized area of the San Juan. The lodes occur at a slightly older geologic horizon than in the other districts and are definitely characterized by formation at moderate depths. The Rico deposits may be considered as on the southern periphery and the Ouray veins as on the northern periphery of the main mineralized region. In these three districts minerals of comparatively shallow formation alone occur. The ores of Silverton and Telluride, on the other hand, may be regarded as having been formed in the heart of the mineralized region where hydrothermal and eruptive activity was most intense, where veins were formed under conditions of high temperature and pressure, and where deposition occurred under heavier cover of overlying volcanics. In these districts a few minerals of the deeper vein zone occur.

ORE DEPOSITS OF LAKE CITY DISTRICT.

THE LODES.

GENERAL CHARACTER.

The lodes of the Lake City area are fissure veins. In the ordinary understanding of the term a fissure vein is a crack or crevice in the rock filled with later-introduced vein material. The Lake City lodes are only in part of this type. They have also been formed largely by replacement and exhibit all stages of the transition from a simple filled fissure, whose walls have been comparatively little affected by vein-forming solutions, to a zone of sheeting and brecciation where most of the mass of vein material has been produced by the alteration of the inclosing rock. Both types of mineralization are common in the same lode, one prevailing in one part, the other in another. In the Golden Wonder lode in Deadman Gulch the fractures are ill defined and the replacement has been extensive and irregular. This is perhaps the one deposit in the Lake City district that may be termed a characteristic replacement as distinguished from a fissure vein.

Ransome[1] has used the term "lode fissure" for those veins whose included mineral has been largely introduced through small, closely spaced fractures, from which solutions have replaced the intervening rock. The Lake City lodes are in the nature of "lode fissures," but, as already stated, so many of them verge toward a simple filled fissure that a clean line of division can not as a rule be made. In some places, as in the Golden Fleece mine, the vein, although it has a very well-defined linear form, consists of a broken zone in which the filling of interstices between rock fragments has been the dominant process, and replacement has been secondary in spite of the extremely broken and shattered character of the zone of mineralization. This vein is discussed more fully on pages 104–111.

Included fragments are present in all of the lodes, and many of the fragments show a very high degree of alteration, but some have

[1] Ransome, F. L., A report on the economic geology of the Silverton quadrangle, Colo.: Bull. U. S. Geol. Survey No. 182, 1901.

been simply surrounded by vein material and have undergone little alteration from the introduction of mineralizing water.

MECHANICAL DISINTEGRATION.

In a region like that at Lake City, where the land surfaces are precipitous, erosion proceeds with great rapidity and the veins have consequently been much dissected. The difference between the lowest and highest topographic points (8,758 and 14,306 above sea level) within the mineralized region is 5,548 feet. The highest point in the outcrop of the several veins examined is 12,800 feet and the deepest point that has been reached in mining on any vein is 7,900 feet. So far as yet determined, therefore, it appears that the veins extend over a vertical range of 5,000 feet. The rarity of mineralized outcrops above an altitude of 11,000 feet renders it probable that their upper limit does not extend much beyond this level. If the length of the Ute-Hidden Treasure vein be regarded as an approximate indication of the depth of the fissure (p. 39), we should have a presumable lower limit of known fissure formation of 7,400 feet.[1] It would therefore appear that the formation of the known Lake City fissures has taken place within a vertical range of 5,400 feet and that almost the entire range of fissure formation is revealed in one place or another by the deep erosion. The mechanical disintegration of the lodes has, moreover, proceeded with great rapidity, for slopes are steep and frost action through a large part of the year relatively intense. Owing to the deep erosion of the country rock in which the veins are contained, great lengths of outcrop have been developed.

As a further consequence of the steep and precipitous nature of the land surface, the veins show fairly well on slopes. Because of the glacial and landslide action, débris has accumulated near the stream levels and has covered the outcrops there, leaving the veins exposed high up on the mountain slopes only. Hence the original discovery shafts on a great many of the properties have been sunk at rather high elevations and carried down to a depth that would ordinarily give the operators an idea of the dip and strike of the lode. The intersection of dipping veins with the steep surface, however, leads to confusion, so that further exploration is generally made by crosscut. Often these crosscuts are unsuccessful, this being due to the fact that after a vein attains any depth in this region it generally undergoes a great change in vein filling and suffers a marked decrease in values. Hence, if cut at all (and often it is not), its appearance is so different that it is seldom recognized as the one that was worked four or five hundred feet above. If the vein were stripped for a vertical distance of several hundred feet down the slope, and drifts run on it, there would be less doubt as to its identity, and as work

[1] Irving, J. D., Ore deposits of the Ouray district, Colo.: Bull. U. S. Geol. Survey No. 260, 1904, p. 59.

progressed the operators might judge whether the work was worth continuing. With everything in sight there would be less dead work and uncertainty, and the cost of stripping the vein would generally be far less than the expense of running a long crosscut.

Where veins have been sufficiently well disclosed to make certain of their continuation in depth their operation by means of crosscuts or drifts has been of distinct advantage, as these have afforded easy drainage and have permitted the development of much stoping ground at moderate cost.

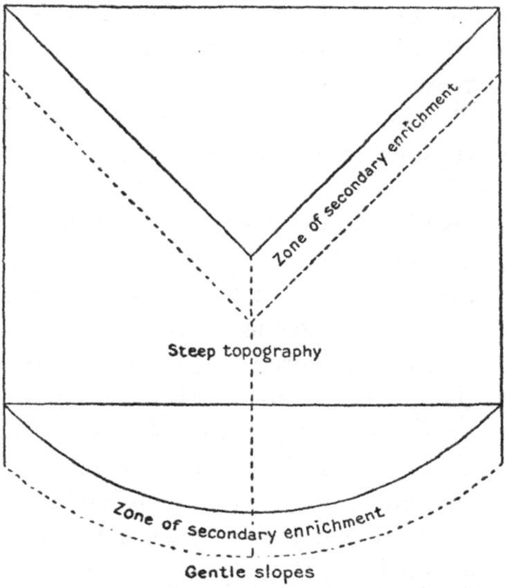

FIGURE 3.—Relation of erosion to superficial alteration.

The steep slopes have not only been of great advantage in exploitation but, in view of the nature of the oxidation and secondary enrichment, have been one of the most valuable assets of the district. If the ores several hundred feet below the surface were as good as those near the upper levels, this would not be so. As conditions are it is most fortunate, for the zone of secondary enrichment is near the surface and nowhere else, and the number of linear feet along the veins is much increased by the steepness of the slopes. Figure 3 shows the relative zones of secondary enrichment for a given horizontal distance in countries with steep and gentle topography. In the latter type there is obviously less enrichment for a given horizontal distance on the vein.

DIMENSIONS.

The following table gives the dimensions of those fissures which have been most satisfactorily explored:

Dimensions of Lake City lodes.

	Length.	Depth.	Width.[a]
	Feet.	*Feet.*	*Feet.*
Lellie	720+	700	0.5-4
Ulay	380+	1,600	5
Ute-Hidden Treasure	2,700+	1,400+	4
Pelican	900+	275+	0.3-4
Missouri Favourite	350+	400+	0.8-2
Nellie M.	700+	500+	2
Monte Queen	950+	600+	3
Black Crook	1,965	1,300	1.5-8
Contention	700+	700	1.5
Golden Fleece	1,300	1,464	0.5-10
Moro-Hendrison	2,000	500	1.2-3

[a] These figures represent rough estimates. They cover vein filling only, not altered or replaced parts of vein walls.

In this region, as in nearly all mining districts, data regarding the dimensions of fissures can be obtained only within the limits of mining operation. Veins are seldom worked to their terminations. The plus sign has been used in the table to indicate that the vein continues beyond the limit of exploration. Where the plus sign has been omitted the work has been continued beyond the point of profitable extraction and has explored the fissure to its apparent termination. The same remarks apply to the depth, although this has been more frequently determined definitely than the extension on the strike.

In length the Lake City veins average between 1,000 and 1,200 feet. Exceptionally strong wide lodes, such as the Ute–Hidden Treasure vein, Black Crook, and Golden Fleece, extend for nearly 3,000 feet. The vertical range of the fissures seems to be about equal to their explored lengths along the strike. The depths in the table given are not the depths of the shafts but the vertical distance between the highest point on the outcrop and the deepest part of the vein below this point. In a great many mines work has ceased before the vein has disappeared, especially where adit tunnels have not been run at lower levels to search for the continuation of the vein. Among the exceptionally long veins is the Moro-Hendrison, which has a length of 2,000 feet, if the correlation of its two parts on either side of the small gulch is correct. It probably represents the root of a fissure, the greater portion of which has been removed by erosion. The Vermont–Ocean Wave–Wave of the Ocean vein is also a vein root. The workings on this vein are now practically inaccessible, so that the writers were unable to determine whether they are all on the same fissure or not, but from the surface outcrops it seems very probable that they are. The vein is fairly straight, for the curves indicated on the map are in large measure due to the effect of topography on the southward dip. If these veins form a single continuous fissure, its total length is over 5,000 feet. The Vermont tunnel, run from the bottom of Henson Creek to tap the vein 1,100 feet below the highest point on the outcrop, has been a failure. Admission to this tunnel could not be secured, but it was reported that the vein was not discovered in the workings, and it is probable that it pinched out above the tunnel level. The Red Rover tunnel, which should have intersected this fissure, disclosed no indication of it. The probability that this vein is a fissure whose upper portions have been completely eroded and whose roots alone remain is strengthened by the extremely rich ores discovered in the upper workings and their very rapid impoverishment with depth. A strong contrast between primary vein filling, such as occurs in the roots of fissures, and secondary enrichment products is, in the writers' opinion, an indisputable proof of the previous existence of a very large vertical range of material from which this rich secondary ore can have been derived and concentrated. In other words, the greater the contrast between the secondary and oxidized ores and the primary ore, the nearer is the approach to the point at which the vein may be expected to disappear completely.

The widths of the veins in the Lake City region vary between a few inches and 20 feet. The average is approximately 18 inches. Many veins were wider in their upper portions and grew gradually narrower with depth. This was the case in the Lellie, Ulay, Black Crook, Golden Fleece, and Vermont. Practically all veins explored by deep workings have pinched out almost entirely. Few widths of 20 feet are found and these extend for short distances only. That in the Hidden Treasure mine was apparently produced by the intersection of a branch vein. Widths of 8 and 10 feet are found in a few places in the Ute and Black Crook veins and according to report in certain portions of the Golden Fleece vein. Pinches and swells in the vein occur both in strike and in depth, and it is, indeed, to these that the division of ore into shoots is chiefly due. Even the Ute vein, which is a singularly uniformly wide fissure throughout its length and probably approaches more nearly to the ideal type of fissure vein than any other in the region, is notably irregular and subject to many pinches and swells in its extension into the Hidden Treasure ground.

TERMINATIONS.

Veins that terminate in depth either narrow into a single small nssure, as in the Lellie, Black Crook, and Golden Fleece, or divide into a number of stringers which finally disappear entirely. Terminations along the strike usually show a division into many branches which finally disappear, as at the southwest extremity of the Ute vein. A sufficient number of examples, however, could not be examined to justify any general rule. As all of the upper terminations of the fissures are now eroded, their extremities in that direction could not be studied.

STRIKE.

Along the strike none of the veins are straight, but twist and turn, generally with sharp angles, somewhat in the manner of a flash of lightning. This is admirably brought out by the plans of the Pelican and other veins. Some of the veins seem to have formed along two intersecting lines of weakness; branch veinlets continue along the old direction, though the main vein assumes a new trend. (See fig. 5.) A marked conformity of jointing and vein direction prevails throughout the district.

Most of the veins in the Lake City area strike in one or the other of two general directions, northeast-southwest and northwest-southeast. The richest lodes trend northeast, but this is probably of little significance so far as the relation of the ore deposits to the geology is concerned. The prominent directions of jointing are approximately the same as those of the fissures. The directions correspond in general to those prevalent in the Silverton quadrangle, although, contrary to natural supposition, the prominent veins in the southwest portion of the Lake City quadrangle do not have the same prevailing direction as those in the immediately adjoining northeast portion of the Silverton, the predominant veins in the former striking northwest-southeast and, according to Ransome,[1] the predominant veins in the latter striking northeast-southwest.

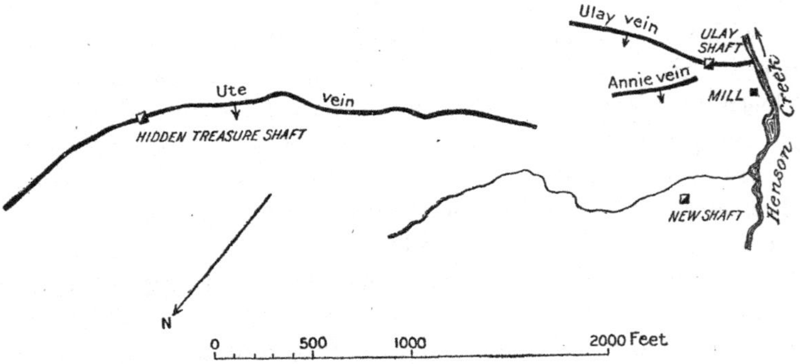

FIGURE 4.—Sketch plan of the Ute-Hidden Treasure group of veins, showing their relations to one another.

DIP.

Nearly all the lodes have steep dips, ranging from 45° to 90°. Only one with a dip less than 50° is known, and that one continues for only a short distance. A few veins are practically vertical. The common inclination is between 60° and 70°. In the Capitol City group of veins the dip is uniformly east, but in the other more widely scattered fissures it varies greatly, dipping here on one side and there on the other. In depth the dips are nearly as irregular. Where there has been much movement this feature has also produced differences in width of the veins along the dip similar to those which occur along the strike.

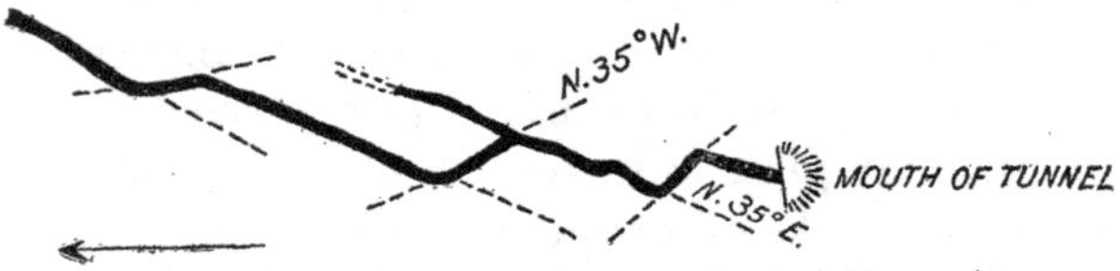

FIGURE 5.—Sketch of typical "forked lightning" fissure vein.

INTERSECTIONS.

Intersections of fissures with different trends undoubtedly take place in many veins, but they can seldom be observed. The Ilma

[1] Ransome, F. L., Economic geology of Silverton quadrangle, Colo.: Bull. U. S. Geol. Survey No. 182, 1901, p. 46.

vein, which runs nearly north and south, intersects the Golden Fleece vein. The actual intersection can not be observed, but it lies on the eastern boundary of the rich ore shoot, which is the most prominent feature of the Golden Fleece vein, and with little question had some effect in producing this ore body.

FAULTS.

Slickensides are common in the Lake City fissures, but they generally indicate movement subsequent to the vein filling. Displacements undoubtedly exist between the two walls of any single fissure, and the large quantities of breccia fragments included in most of the fissures, the prevalence of pinches and swells in the veins, and the selvage clays commonly noted point to some movement between vein walls. The extent of this can not be determined, as there are no recognizable beds in the alternating complex of volcanics to serve as a basis of measurement. It is believed, however, that the faulting along fissures has in general been comparatively slight.

Few of the lodes are disturbed by later movement. A definite fault was observed in the Ilma vein which displaces the vein 35 feet, and post-mineral faulting was observed in the Gallic ore body.

ORIGIN OF THE LODES.

Too few fissures have been explored in the Lake City country to permit any generalization as to their origin. It seems probable that they were produced by the same causes that gave rise to the fissures in Silverton, Ouray, and Telluride. These causes were undoubtedly operative subsequent to the invasion of the volcanic series by the monzonite masses, for some of the fissures cut this rock. It is believed that the fissures were produced by compressive strains due possibly to the gravitative readjustment which accompanied recent movements in the region.

In general, the writers believe that there is no evidence for attributing different ages of formation to the lodes which have different trends. They are believed to have been formed during a single period of fissure formation and to have been mineralized also during a single period.

DISTRIBUTION OF THE LODES.

Most of the veins of the Lake City mining district are located on the slopes of gulches which drain into Henson Creek. A few are on the slope north of the Lake Fork of the Gunnison. Of these two localities the former is a part of the Lake City quadrangle, and the latter is so near it and of such historic interest that it has been thought advisable to incorporate it in the report on this area.

The veins shown on the map (Pl. I) are those that have been worked, and they were in all probability the most easily discoverable in the district. The writers see no geologic reason why the veins should be so segregated as the map indicates, and they are inclined to think that others exist which may prove equally valuable if discovered. In other words, the area is one in which local conditions have not been the cause of the formation of fissures and the subsequent filling with vein material. The conditions were widespread and it is only natural to suppose that the processes which formed one vein have been equally active in other parts of the mining district proper.

FISSURE FILLING AND METASOMATISM.

Mineral solutions which have penetrated the fissures have deposited gangue and ore minerals in them and have altered the adjoining rock. In very few fissures is alteration of the wall rock entirely lacking, and in some of them (pp. 113–114) it has been so extreme that the entire ore mass in the mine is to be attributed to it. In general, however, silicification and sericitization have been the predominant types of alteration. These have resulted in a very fine-grained dense-black jasperoid material which in places extends 4 or 5 feet from the vein filling. Fragments of included rock have been especially subjected to this type of alteration and are usually spoken of in the mines as black quartz. A later alteration of this black material is not uncommon and a light-green margin has been developed extending from a fraction of an inch to 6 inches from the vein filling into the previously altered country rock. Many included fragments have been completely altered to this light-green material, but small cores having the shape of the fragment are not uncommonly left in the center. Microscopic study of the black material shows that it is composed of very finely divided secondary silica and a great abundance of extremely minute particles of sericite. The greenish material is generally marked by the coarser crystallization of the quartz and a relatively smaller quantity of sericite. The original rock is either an andesite breccia or a solid andesite with glassy or cryptocrystalline groundmass. Barite, rhodochrosite, sphalerite, galena, and tetrahedrite replacing the wall rock beyond the limits of the open spaces have not been observed by the writers. Their occurrence, however, as distinct crystals in the black silicified fragments shows that they have been deposited either as replacements of an already altered country rock or have replaced these fragments previous to their silicification. Pyrite, on the other hand, commonly extends into the country rock farther than even the silicification and sericitization. It is then well crystallized into minute cubes. Studies of the paragenesis of the ores indicate that the pyrite, silica, and sericite represent the earlier stages of vein formation and are probably the first results of solutions entering the fissures.

BANDING.

Where fissures have been filled with ore minerals or where included plates of country rock have been completely replaced, banded structure is commonly well developed. This is the case in the ore from the Hidden Treasure, also in the ore from the new shaft of the Ulay mine. The ore in the Ute vein is also roughly banded. No specimen observed, however, shows well-developed comb structure throughout the vein. Almost universally one band grades into another, making it in many places impossible to determine the relative ages of formation. It may be said, however, that white crystalline quartz is as a rule the latest mineral deposited. It commonly cements shattered galena and sphalerite or chalcopyrite, rhodochrosite, and tetrahedrite. It also coats crystals of barite which project into cavities. The tetrahedrite in veins belonging to the tetrahedrite-rhodochrosite group (see p. 47), is later in formation than the galena, as it permeates that mineral in many places along cracks and fractures. It is generally closely associated with the rhodochrosite, being commonly scattered through the rhodochrosite mass without any regularity. In general, silicification and sericitization of the country rock was followed by the deposition of vein minerals in the following order: (1) Pyrite in the wall rock and in the fissure; (2) rhodochrosite-galena-sphalerite; (3) tetrahedrite; (4) white crystalline quartz; (5) secondary sulphide-enrichment minerals. In many places, however, the sphalerite, galena, and rhodochrosite show reversals in the order of their formation, and in many it is difficult if not impossible to determine their relative ages with any certainty.

MINERALIZATION.

AGE.

The veins of the Lake City region cut all flow rocks except the Potosi volcanic series and the still later rhyolites and basalts. They cut even the monzonite porphyry intrusions, which are believed to be of late Eocene age. The vulcan vein cut in the Gallic tunnel passes through both monzonite porphyry and andesite, and the Chord Extension vein occupies a fissure entirely in monzonite porphyry, both furnishing conclusive evidence that the fissuring has taken place since the latest injection of magma. Further exploration in the Lake City district may reveal veins cutting the Potosi lavas; in the northern part of the Telluride quadrangle, the veins do cut the Potosi and are mineralized therein, though perhaps not so frequently as in the lower and older rocks; and some of the Silverton lodes cut the Potosi. If the mineralization took place at approximately the same period in these three regions, it is evidently post-Potosi; that is, late Miocene or early Pliocene in age. Ransome believes that the formation of the veins in the Silverton region extended even into the Pliocene. However, there is no way of definitely proving that the veins were formed contemporaneously, and so no fixed age can be given to the period of vein formation and mineralization.

Some connection may exist between the fissuring of the region and the intrusion of the porphyry; for instance, the fissures may have relieved the strains caused by the intrusives. Besides the first great period of fissuring there have been at least two subsequent periods of displacement. In the Gallic mine there is evidence of a primary deposition of ore and gangue minerals, followed by faulting and fissuring, with brecciation of the country rock and vein material, which was later recemented by a second deposition of quartz carrying various ore minerals. A third period of movement is shown by the seams of gouge which are present in different parts of the mine. The Woodstock prospect in Yellowstone Gulch shows evidences of a brecciated vein recemented by a second deposition of quartz. Probably all of the disturbances which have occurred since the original veins were formed have been of a minor character.

EFFECT OF COUNTRY ROCK.

The effect of the country rock upon the vein filling has apparently been of no marked consequence for the minerals in many of the veins vary at places where no change occurs in the wall rock and, on the other hand, veins whose mineralogy is the same throughout occur in all of the rock formations. The Hidden Treasure mine, located in andesite of Picayune volcanic group, has a gangue of rhodochrosite, barite, quartz, and fluorite. The Ute, a continuation of the same vein, has quartz and barite. The prospects in and around Owl Gulch, which are in the same formation as the Ute and Hidden Treasure, have mainly quartz. Mines in Eureka rhyolite in Yellowstone Gulch have quartz as the predominant vein filling, but the Pelican, in the same formation east of Sugarloaf Rock, has barite. Barite is also the prevalent gangue mineral in the Missouri Favorite. Evidently the nature of the country rock was not the dominant factor in the deposition of the vein minerals.

RELATIVE ABUNDANCE OF MINERALS.

MINERAL GROUPS.

A preliminary idea of the mineralogy of the Lake City district has been given in Table 1 (p. 34), which was compiled to show the mineral relations among the districts of the San Juan region. More detailed data are given in Table 2 (opposite), in which the relative abundance of the minerals in the mines is shown by the size and character of the type. The list of minerals has been made as complete as the conditions would permit, but it is probable that it would have shown much greater variety if all parts of the lodes had been accessible. Some errors may also have crept in through inaccurate identification, but in general the list is believed to be fairly accurate.

It is impossible to divide the lodes into groups separated by absolutely sharp and characteristic mineralogical differences. They may, however, be divided, according to the relative abundance of their component minerals, into three fairly distinct types, which, designated by their most distinctive minerals, are the tetrahedrite-rhodochrosite group, the quartz-galena-sphalerite group, and the telluride group. These groups, especially the first two, merge into one another at many localities and are without doubt due rather to local variations in the nature of solutions than to separate periods of mineralization.

TETRAHEDRITE-RHODOCHROSITE GROUP.

The tetrahedrite-rhodochrosite group of veins comprises those whose ores consist of dominant galena and argentiferous tetrahedrite with considerable sphalerite and some pyrite in a gangue composed chiefly of quartz, rhodochrosite, and barite. The distinctive or diagnostic minerals in this group of veins are preponderating tetrahedrite, rhodochrosite, and abundant barite. Pyrite is usually subordinate in quantity. Chalcopyrite is also subordinate. Gold values are invariably low and the veins produce chiefly lead and silver. Copper is a by-product. Zinc is not generally present in paying quantities and unless saved in the mill is objectionable. The unenriched primary ore in these veins varies in silver value in accordance with the proportion of silver-bearing tetrahedrite which is present. The gold seems more closely associated with the pyrite than any other mineral. Local increases in chalcopyrite sometimes render the copper values important. The mines whose ores fall into this class are the following:

Hidden Treasure.
Missouri Favorite.
Ute and Ulay.
Pride of America.
Casino.
Lellie.
Vermont.
Ocean Wave.
Wave of the Ocean.
Black Crook.
Belle of the West (?).
Contention.
Silver Chord Extension.

QUARTZ-GALENA-SPHALERITE GROUP.

The veins of the quartz-galena-sphalerite group are characterized by dominant galena and sphalerite with usually subordinate chalcopyrite in a quartz gangue. Barite is either absent entirely or very subordinate in quantity. Tetrahedrite is present in many mines, but is not prominent. It is, however, sufficiently abundant to yield with the argentiferous galena, the rich secondary minerals which have enabled the mines to produce silver as their most important product. Gold values are a little more important than in veins of the tetrahedrite-rhodochrosite type and seem to be associated with the larger quantities of pyrite present in the veins of this group.

Copper and lead are both important products, and in some mines the sphalerite itself is sufficiently abundant to yield profitable returns.

Two or three of the mines of the tetrahedrite-rhodochrosite and quartz-galena-sphalerite groups show transitions one toward another. The Silver Chord Extension mine, though located in the midst of the series of coordinate lodes belonging to the quartz-galena-sphalerite type, shows all of the characteristics of the tetrahedrite series. Again, the Ute vein, the most productive vein in the region, shows affinities toward the quartz-galena-sphalerite type in the Ute ground, but farther north in the Hidden Treasure ground is distinctly a member of the tetrahedrite series. In other mines some shoots show affinities toward one group and other shoots in the same vein show affinities toward the other group. It is obvious that no sharp line of demarcation exists between the two groups, and it is therefore probable that they belong to a single period of mineralization and were contemporaneous in their origin.

TELLURIDE GROUP.

The telluride group consists of veins containing tellurides disseminated through a fine-grained quartz gangue (Golden Fleece), with subordinate galena, sphalerite, pyrite, chalcopyrite, tetrahedrite, hinsdalite and barite. If it were not for the tellurides it would be impossible to distinguish these veins from those of the normal tetrahedrite group; but as the tellurides are entirely absent in the other vein types, they set the former sharply apart. The presence in the Golden Fleece of hinsdalite associated closely with tellurides gives to this vein a somewhat unique character. Neither the telluride nor the hinsdalite are, in the writers' opinion, sufficient to indicate that the Golden Fleece vein has an origin different from the other veins or belongs to a separate period of mineralization. It rather seems to be a variation from the normal type, such as may be frequently encountered in almost any connected area of mineralization. The products of the telluride group are both silver and gold. In the Gallic-Vulcan mine the silver would probably predominate if the mine were of productive size. In the Golden Fleece the proportion by value of gold and silver in the ore was approximately 1:1 (see pp. 110–111), a very much higher proportion of gold than is characteristic of any of the other mines.

SOURCE OF MINERALIZATION.

It is hardly within the province of a brief paper dealing with a small portion of an extensive mineralized region such as the San Juan to enter into an extended discussion of the problems connected with the origin of the ores. Both Ransome [1] and Purington [2] have dis-

[1] Bull. U. S. Geol. Survey No. 182, 1901, pp. 132–141.

[2] Eighteenth Ann. Rept. U. S. Geol. Survey, pt. 3, 1898, pp. 819–824.

cussed the genesis of the San Juan lodes at considerable length, and inasmuch as the Lake City lodes are probably of the same origin a further discussion may seem redundant. Nevertheless, since later study of genetic problems has led to results applicable to the Lake City and the other related San Juan lodes, some suggestive thoughts seem worthy of presentation in connection with this question.

The explanations offered to account for the filling of mineral veins have of late centered around two opposing hypotheses; the first holds that meteoric waters have dissolved out minute quantities of metallic and nonmetallic elements from the country rocks, have descended into regions of elevated temperature, become heated, and have risen up through fissures and deposited their burden in mineral veins. The second holds that the metals and the accompanying elements and moisture have been component parts of deep-seated eruptive magmas and have, as a final stage of the cooling of the magmas, been released, have found their way upward and have been deposited as ores in mineral veins. A compromise between the two views has often been advocated in the form of a mingling of waters and contained metals derived from both sources.

To the second theory, that of derivation from eruptive rocks, the name "magmatic" origin is now generally applied. The older term "pneumatolytic" was formerly commonly applied to ores supposed to have this derivation, but was used in a much narrower sense, being applied only to deposits like the Cornwall tin veins, where minerals rich in boron, fluorine, and tin were important constituents, and where the veins were distributed in the immediate vicinity of large deep-seated masses of intrusive igneous rocks.

In his discussion of the San Juan lodes Ransome considers both of these possible modes of origin and dismisses the pneumatolytic as improbable. The following paragraph seems to fairly well epitomize his views:

The Silverton lodes, as a whole, possess neither the distribution nor the mineralogical characters of those deposits to which, in the light of present knowledge, an essentially pneumatolytic origin can be most safely assigned. The known igneous masses had certainly solidified and probably had lost much of their heat before the lode fissures were formed. It is most probable that the transportation and concentration of the Silverton ores was effected chiefly by meteoric waters, which derived their chemical and mechanical energy mainly from the heat connected with volcanism and from pressure, but possibly in some minor part also from gases and vapors given off at high temperatures by solidifying igneous rocks and taken into the deeper meteoric circulation.

There can be no question that the San Juan lodes possess none of the characters of pneumatolytic deposits in the older and narrower sense of the term either in their mineralogy or in their distribution, by which last is probably meant their occurrence as an aureole of metalliferous deposits about a deep-seated igneous mass. As stated above, however, such a character is not an essential feature of the magmatic origin as now understood.

The minerals formed by any given set of vapors, as the writers understand the question, depend on the temperatures and pressures under which deposition has occurred; and such temperatures and pressures, in their turn, are generally a measure of the distances to which the vapors have migrated from the parent magma since their emission. In other words, the absence of minerals characteristic of pneumatolytic deposits from the San Juan lodes does not signify that these lodes were not deposited from magmatic waters, but simply that those waters had migrated to a zone where temperatures and pressures were so low that the so-called pneumatolytic minerals could not form, because they were unstable compounds under those physical conditions.

The comparative mineralogical table (p. 34) shows that in the adjoining Telluride region a number of minerals characteristic of high temperatures and pressures have actually been deposited; hence in that portion of the San Juan region the requisite conditions of temperature and pressure did in some small measure actually obtain. It seems, therefore, that the mineralogy of the lodes offers no obstacle to this explanation of their derivation from magmatic waters.

The absence of any extensive eruptive masses, however, from which magmatic waters may have had their origin does apparently offer a much more serious difficulty.

Ransome's studies in Silverton, like those of the writers in Ouray and Lake City, show conclusively that the veins, and hence also the contained minerals, cut the monzonites and monzonite porphyries, which, aside from the rhyolites and basalts of post-Potosi age (apparently barren of any mineralization), are the latest manifestation of eruptive activity in the region. The veins are therefore regarded by Ransome as later than the monzonite porphyries and consequently as independent of them in their origin. Hence if the minerals are derived from an igneous source, that source must be some deep mass of later age concerning the presence of which there is no geologic proof.

If the monzonite and monzonite porphyry bodies which cut the extrusive volcanics in the Silverton, Ouray, Lake City, and Telluride regions, and which are in fact widely prevalent intrusive rocks throughout the entire San Juan region, did not precede the fissure formation, they could be regarded with confidence as the igneous rock which has had most to do, not only with these ores but with the others which are so prevalent throughout this mineralized area. This is true because they represent the latest intrusives and because they have not escaped to the surface carrying their vapors with them, but have cooled at depths and have yielded mineralizing waters as a final phase of their consolidation.

To the writers the intersection of the upper and first-cooled portions of the monzonite masses by the fissures does not in the least preclude the deeper portions of the monzonite magma from consideration as the most probable source of these ores. Monzonite, quartz monzonite, granite, and quartz diorite have been so often observed by the writers in association with metalliferous ores that their efficiency as a cause seems to them certainly well established. Moreover, as the emission of gases and vapors is the final phase of consolidation, the outer portions must have cooled sufficiently to permit fracture and venation before the lower portions could have yielded their mineralizers. A close analogue can be seen in this very region in the flow breccias, in which fragments of andesite may be seen embedded in andesite. Both fragment and matrix are portions of the same magma; the fragment merely represents the outer portion cooled first and later ruptured by lava movements and then cemented by still molten andesite. In like manner with the veins, except that here, instead of molten rock and small fragments, emitted vapors and widely separated fractures must be considered. Instances of this kind are too frequent to need extended comment. In the Black Hills of South Dakota the refractory siliceous ores are later in age than the phonolite rocks which represent the latest phase of igneous activity, but from this alkali-rich magma these ores are almost certainly derived.

The potency of the monzonite magma to produce mineral deposits is amply attested by the contact deposits that it has produced in the limestone masses in Ouray and, as shown by Ransome's description, in Rico. Unquestionably these contact deposits with their characteristic minerals are derived from emissions from the magmas in situ. The solidification of the main mass of rock and the final emission of vapors which migrated upward into the fissures, some of which had cut their cooler upper portions, seems to the writers to represent by far the most probable sequence of events by which the San Juan lodes have been produced.

MINERALOGY.

The mineral species of the Lake City district are either primary or secondary. The primary minerals are (1) minerals formed at shallow or moderate depths and (2) persistent minerals common to all depths. The secondary minerals are (1) minerals due to oxidation processes and (2) minerals that result from secondary sulphide enrichment. No minerals characteristic of the deeper zones appear in the Lake City lodes, and (p. 36) it is therefore probable that the latter were formed at moderate depths below the surface. In other words, the covering of superincumbent rock has been lighter in this region than in the adjacent Silverton and Telluride districts.

In the following pages the minerals of shallow and moderate depths are first considered; then those which may be called persistent minerals. Minerals of surficial origin, including oxidation products and minerals due to secondary enrichment, are taken up last.

PRIMARY MINERALS.

MINERALS FORMED AT MODERATE AND SHALLOW DEPTHS.

Tetrahedrite.—Tetrahedrite ($Cu_8Sb_2S_7$) is one of the most prevalent and, because of its silver content, one of the most important ore minerals in the Lake City country. In many places it is conspicuous in the ore, but even where it can be detected with less ease it is very important. In greater or less quantity it is present in all mines of the district and, in fact, is reported in some quantity in nearly all mines of the San Juan region.

Ordinary tetrahedrite of the formula given is usually designated "fahlerz" by the Germans, but this type of the mineral does not predominate in this Lake City region. The copper is almost everywhere partly replaced by silver and the antimony to some extent by arsenic. The following analysis from Genth[1] shows the composition of a nonargentiferous tetrahedrite:

Analysis of tetrahedrite.

Sulphur (S)	25.97
Antimony (Sb)	25.51
Arsenic (As)	3.22
Copper (Cu)	37.68
Iron (Fe)	.64
Zinc (Zn)	7.15
Silver (Ag)	.60
Bismuth (Bi)	.37
Manganese (Mn)	.10

Typical freibergite from Freiberg contains more than 30 per cent of silver. The freibergite found in the Lake City quadrangle carries large proportions of silver, much of it running $200 to $300 to the ton, and some that is nearly pure reaching even 2,500 ounces to the ton. The variety that contains much silver has usually a light steel-gray color and is somewhat more greasy in appearance than the nonargentiferous variety; also, the streak which in normal varieties is gray to black is somewhat reddish in the argentiferous types.

The mineral in these lodes is invariably massive, never in crystals, is commonly mingled intimately with galena, and, in general, is highly argentiferous. The correlative mineral tennantite has not been recognized, and the arsenic is probably not commonly great in amount, for almost all the ores yield pyragyrite and not proustite on secondary alteration.

[1] Genth, F. A., Proc. Am. Philos. Soc., vol. 23, 1885, p. 38.

Bismuth compounds.—The complex sulphur compounds of bismuth are reported in considerable quantities in the Monte Queen mine. A complete analysis of this mineral is not available, so that its exact character is not known. According to the operators it contains high percentages of silver, about 20 per cent of bismuth, and considerable zinc. It is gray in color, resembling tetrahedrite, but containing little or no copper. It seems not unlikely that it results from secondary sulphide enrichment, for it lies close to the oxidized zone and is said to carry much higher percentages of silver than most of the other ore in the Monte Queen mine. Further study, however, is necessary to determine its chemical character and origin.

Barite.—Barite ($BaSO_4$) is abundant in the gangue of the galena-sphalerite veins, especially in those of the variety carrying tetrahedrite, which are developed most characteristically along Henson Creek and near Lake San Cristobal, but is less common in those of the Capitol City type, in which quartz predominates. It is present also in much smaller quantity in the telluride veins, where silica seems to be the predominant gangue. The barite as a rule is an interlocking network of thin plates whose interstices are filled by fine-grained silica or metallic minerals. This structure discloses but little banding and gives a massive appearance to the ore. In many vugs it is developed in very beautiful crystals of considerable size and generally perfect transparency. In a few veins it exceeds all other minerals in quantity, and when this is the case the ore is of little value. The barite is clearly of earlier deposition than the quartz, as the latter frequently incrusts crystals that project into central cavities. Barite is very abundant in the Hidden Treasure portion of the Ute-Hidden Treasure vein. It offers a strong contrast to the quartz in most of the lodes where it is present. Perhaps its commonest association is with the jasperoid or fine-grained quartz described on pages 44 and 60.

Rhodochrosite.—Rhodochrosite ($MnCO_3$) occurs in many of the veins in the Lake City region and in some lodes is absent in one portion and present in great quantity in another. Thus in the Hidden Treasure mine it makes up the bulk of the vein filling in the northern end of the Hidden Treasure ground and is practically absent in the Ute end of the same vein. Again, in the Monte Queen mine it is inconspicuous in most of the stopes but forms a full 5-foot face of solid mineral in the extreme western face of the tunnel. Many of the stopes in the Ilma vein likewise show no evidence of rhodochrosite, though in others it is present in large quantities and is notable for its coarsely crystalline structure and deep red color.

The majority of the rhodochrosite in all of these veins is contained in a very fine-grained aggregate whose individual cleavage faces measure not over 1.5 millimeters. When first mined it is generally deep pink in color but rapidly bleaches on exposure until it has only a slight pinkish tinge which distinguishes it from dolomite. If the exposure be long continued it develops a brownish coating, which makes it resemble siderite unless examined on fresh fracture. Rhombohedral crystals of rhodochrosite are common in cavities, in some of which they are intergrown with quartz. It is probably one of the later minerals deposited. Tetrahedrite is more generally associated with rhodochrosite than with any of the other minerals; so much so that in milling the crushed ore on the Wilfley tables in the Hidden Treasure mill about an inch of rhodochrosite above the line of concentrates and gangue is saved from the tables. This is done because the rhodochrosite contains considerable quantities of silver even if the included particles of tetrahedrite are too fine for observation.

Hinsdalite.—Hinsdalite ($2(PbSr)O.3Al_2O_3.P_2O_5.2SO_3.6H_2O$) was first collected by E. S. Larsen, to whom belongs the credit of its discovery and investigation. Mr. Larsen will soon publish a longer description of it elsewhere. It was first found on the dump at the mouth of one of the tunnels of the Golden Fleece mine, at an elevation of about 9,950 feet, where it is present in considerable amount. It is an

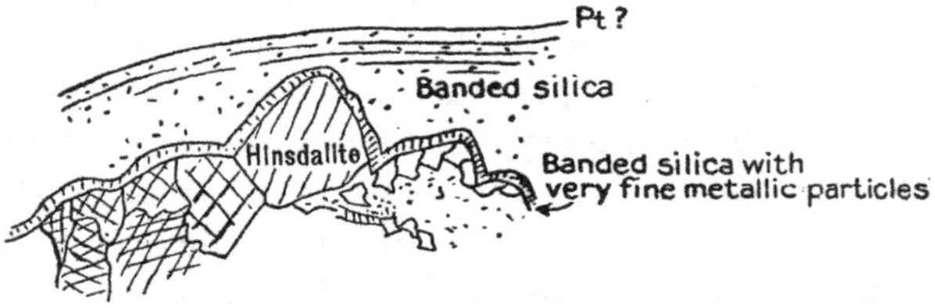

FIGURE 6.—Hinsdalite from the Golden Fleece mine.

original vein mineral associated with quartz and a little pyrite, galenite, tetrahedrite, and barite. It occurs in bands about an inch wide or as crystals imbedded in the grayish chalcedonic quartz which forms the matrix of the petzite in the richest ore. Its crystals are not uniformly well defined as to boundaries, but generally form irregular bodies ranging from minute particles to masses three-fourths of an inch in greatest diameter. The crystals are marked by a very distinct and brilliant cleavage. (See fig. 6.) They are either rhombohedrons, resembling cubes, or pseudohexagonal tablets. If the apparent hexagonal base be taken as the true base, there is a perfect basal cleavage, but the cleavage faces are as a rule wavy and striated. The optical data indicate that the mineral is only pseudohexagonal. Its hardness is about 5; its luster is vitreous to greasy. The fresh mineral is pale greenish, but much of the material is dark gray from inclusions. The streak is colorless.

The indices of refraction are somewhat variable, but the values for the principal zones are about $\alpha=1.670$, $\beta=1.671$, $\gamma=1.688$. Sections normal to the cleavage show parallel extinction, those parallel to the cleavage are generally hexagonal in outline and show the emergence of the positive acute bisectrix. They may easily be taken for uniaxial crystals, as the axial angle is small but variable; 2 E is usually about 32°. Basal sections are generally divided into six radial segments, and the plane of the optic axis in each segment is normal to the pseudohexagonal prism edge.

W. T. Schaller furnished the following analysis, which was made on fresh light-gray crystals of hinsdalite. They were examined microscopically and found to be very pure, but showed a slight zonal growth.

PbO	31.75
SrO	3.11
Al_2O_3	26.47
SO_3	14.13
P_2O_5	14.50
H_2O	10.25
	100.21

CaO, MgO, Na_2O, K_2O, trace.
D=3.64.
Formula $2(Pb\ Sr)O.3Al_2O_3.P_2O_5.2SO_3.6H_2O$.

Hinsdalite is infusible but whitens on heating. It reacts for aluminum when heated with cobalt nitrate and readily yields a button of metallic lead. It is insoluble in acids. The water is driven off only at a temperature of from about 400° to 600° C.

MINERALS FORMED AT ALL DEPTHS.

Pyrite.—Pyrite (FeS_2) is present in all lodes of the region and is by far the most widely distributed of the metal-bearing minerals. It differs in abundance in the different lodes and is generally found in greatest amount in the lower and less valuable portions of the mines. When associated with galena it is more commonly fine grained and shows few crystal faces. It is subordinate in most of the ore in many lodes, especially in those which consist chiefly of galena and tetrahedrite and those which carry tellurides. When galena is in large amount and quartz is the gangue, pyrite is generally insignificant, as in that portion of the Ute vein which contains chiefly argentiferous galena and sphalerite. In the Capitol City series of galena-sphalerite-chalcopyrite ores it is, however, relatively abundant and in some ores is very conspicuous.

Where not contained in tellurides, gold is apparently more generally contained in and associated with pyrite than with any of the other minerals. In this respect the Lake City pyrite closely resembles that described by Purington from Telluride. In the Contention mine the rich silver values contained in the gray copper and its oxidation products contained but little gold, but in the lower levels the vein carried chiefly quartz and pyrite, and in this mixture the gold was commonly notable, yielding a much better average than in the more profitable silver ores above. Similar conditions were found in the Moro mine.

Pyrite is possibly an exception to the general rule (p. 44) that the metallic minerals are not generally present in the country rock, despite the fact that silica in many places penetrates to great distances and transforms the rock into jasperoid, in which phenocrysts and structures of the original are preserved. In places, as in the lower levels of the Contention mine, pyrite replaces the porphyry and occurs as disseminated crystals, few of which measure more than one-fourth inch and nearly all of which are rather perfect cubes, though generally containing pyritohedral striations. Silica and pyrite, singly or together, seem to have had a permeating power strikingly lacking in the other ingredients of the ore.

Coarsely crystalline pyrite is not common in the Lake City veins, nor are large, massive bodies of it seen in any of the veins.

Galena.—Galena (PbS) is absent in few Lake City veins, but its amount varies greatly in the different lodes and in the different portions of any single lode. In some mines, as the Ute-Hidden Treasure, it forms an extremely abundant and very profitable mineral at one end of the lode, and at the other it sinks greatly in amount and is less abundant than tetrahedrite. In the lead-zinc-copper group of Capitol City it shows the most common association with sphalerite, but in the other mines this association is either lacking or is very much less marked. When not associated with sphalerite, it is commonly intimately mixed with tetrahedrite. The mixture can be detected with the naked eye in many specimens, in which the tetrahedrite seems to fill the interstices between shattered fragments of galena (Pride of America, Missouri Favorite, Pelican, and many others). When not apparent to the eye, its presence may be detected by polishing the surface of a mass of rich silver-bearing galena. Indeed, in the writers' opinion, all the high-grade unenriched galena owes its high silver content to tetrahedrite. Little pure galena in the Lake City lodes carries more than 22 ounces in silver to the ton and much of it carries a good deal less. Similar conditions in the Silverton district are recognized by Ransome, who states[1] that the galena there, when free from the richer silver minerals, does not contain very much silver.

The authors polished carefully some faces of normal Lake City galena carrying about 10 to 15 ounces of silver, but were unable to detect any mechanically mixed mineral in it. In this respect the Lake City mineral offers a parallel to some described from England by Finlayson,[2] who found native silver mechanically mixed with much

---

[1] Ransome, F. L., A report on the economic geology of the Silverton quadrangle, Colo.: Bull. U. S. Geol. Survey No. 182, 1901, p. 80.

[2] Finlayson, A. M., Ore deposition in lead and zinc veins of Great Britain: Quart. Jour. Geol. Soc. London, vol. 66, 1910, p. 319.

of the silver-bearing galena carrying high values, but could find no trace even under higher powers of mechanically mixed minerals in galena carrying lower values. His inference seems justified that the silver to a certain amount is chemically combined in some manner with lead and sulphur, but that when present in very large amount it exists as intermingled free silver. The difference in the two cases is simply that in the one instance it is native silver and in the other argentiferous tetrahedrite. Much of the galena shows complex twinning, due to crushing stress, and this shows admirably in many polished sections. In most of the lodes the galena is rather coarse, but much of the finer variety, even down to "steel galena," is also found. When associated with sphalerite both are either coarsely crystalline, as in the Ute vein, or are extremely fine-grained, as in the Monte Queen.

Some interesting practical results follow from the study of the high-grade silver-bearing galena. The Lake City ores offer considerable difficulty in milling operations (p. 72), and in the Ute mine especially the greatest difficulty was experienced in recovering a reasonably high proportion of silver. The difficulty in saving the silver in the argentiferous galena is without doubt due to the intermixed tetrahedrite, for this mineral is so easily slimed that it floats away on top of the water, and only canvas tables, of which the writers have seen none in the Lake City region, will save any considerable portion of it. With it goes much of the silver, leaving only the chemically combined silver in the galena.

Fully faceted crystals of galena are commonly seen in vugs in the Capitol City and other ores, but few of them attain notable size.

Sphalerite (*zinc blende*).—Next to pyrite, zinc blende (ZnS) is the most abundant and most universally distributed mineral in the lodes of the Lake City region. In practically no mines is it entirely lacking, although it is much more abundant in some than in others. Thus in the highly zinciferous lodes of the Capitol City region it is generally present in such large quantities that the endeavor is now being made to produce a concentrate which will run 29 per cent or more in zinc, and thus enable the companies to secure payment for this metal and avoid the high penalties charged by the smelters for ores which run between 10 and 29 per cent in zinc. In much of the gold-bearing telluride ore of the Golden Fleece sphalerite is absent and where present is extremely subordinate in quantity. The telluride ores in the Gallic-Vulcan mine carry notable amounts of extremely dark-colored ferruginous sphalerite. The mineral in this particular mine, as will be later explained, has exercised considerable effect on the secondary precipitation of gold. In the Capitol City lodes, where sphalerite is present in greatest abundance, it is coarsely crystalline and is usually light brown to light green and even pure yellow. Less commonly it has the dark ferruginous character of the iron-bearing types known as "Black Jack." All varieties, however, are present to some extent in most of the mines of the region. The very fine grained, almost massive type is uncommon in the Lake City ores; it does occur, however, notably in the Monte Queen mine, in which the sphalerite is of unusual interest. It is extremely fine grained, approaching the types known from the Friedensville mines in Pennsylvania, is very dark in color, and carries high values of silver and generally notable percentages of bismuth. It has not been possible to determine from the specimens whether the bismuth is present as an individual mineral species or is in some manner combined with the zinc. Certain analyses of much of this ore are reported to show 20 per cent bismuth. High proportions of bismuth are invariably accompanied by increased proportions of silver, and it seems probable that the ore includes a mineral (perhaps a variety of sphalerite) containing both bismuth and silver in chemical combination. In the Ute vein and the other veins of the district the sphalerite is most commonly associated with galena.

In general the sphalerite belongs to one of the earlier periods of mineral deposition. This is noticeable in the Ute vein and the Moro vein, in both of which the sphalerite is much shattered and penetrated by the white quartz which forms a larger portion of the gangue material. Where zinc is prominent and yet is not sufficiently abundant to be saved it becomes very objectionable and in not a few instances has led to the abandonment of workings.

The vertical range of sphalerite seems to be from the lowest workings in decreasing quantities toward the surface. In the Ute mine all the veins carry a very large increase in this mineral with depth. The Ulay vein, the Ute vein, and the new vein recently opened in the west shaft contain large quantities of zinc and show a decrease in silver content so great that the ore can not be handled profitably. The same is true of some of the Capitol City mines.

Chalcopyrite.—Chalcopyrite ($CuFeS_2$) is especially abundant in veins that carry neither the tellurides nor notable quantities of tetrahedrite. In subordinate amounts it is present in all the mines. It is always massive, never crystalline, and does not generally carry appreciable quantities of either gold or silver. In the Moro mine and the related veins near Capitol City it is especially abundant. In the Henson Creek mines and those in the vicinity of Lake Fork it is subordinate. Where present in large quantities and intimately mingled with sphalerite the separation of the two minerals in milling operations has been attended with serious difficulty, the specific gravities of the two minerals (sphalerite 3.9–4.2, chalcopyrite 4.1–4.3) being so nearly the same that it is extremely difficult to secure a clean concentrate. For this reason the Moro mine has installed a

Blake-Morscher static electric separator. It is not known whether this has proved successful or not.

Chalcopyrite shares with tetrahedrite the copper production of the district, but copper is in all of the lodes essentially a by-product. Tetrahedrite (gray copper ore), where unaccompanied by silver values, is not an important ore in this district, and it is highly improbable that these veins could have been worked for their copper content alone.

Tellurides.—Tellurides are absent in all except two of the Lake City lodes, the Gallic-Vulcan mine and the Golden Fleece mine. The Gallic-Vulcan has never been productive and interest in it is purely scientific, but the very large proportion of tellurides in the Golden Fleece mine gives to the mineral a high relative importance. It is noteworthy that in Rico, Telluride, and Ouray tellurium compounds are entirely absent and in Silverton they are present only as scientific rarities. Their appearance, therefore, in one or two veins in such large quantities and in the midst of other types of minerals gives to Lake City a feature which serves to distinguish it from the other districts of the San Juan region. Tellurides are also reported to occur in the Isolde mine in the Burrows Park region, but these statements have not been verified, as that mine is outside of the area covered by the present examination.

In the Gallic-Vulcan mines the tellurides occur in small quantities distributed through the vugs in the white quartz which constitutes the larger portion of the vein material. They range in color all the way from lemon yellow through greenish yellow to silver white and steel gray. An insufficient amount of pure material was available for analyses, so that the individual species can not be definitely stated, but from their color and general appearance it is probable that all of the varieties calaverite, sylvanite, krennerite, petzite, and hessite are present.

In the Golden Fleece mine the prevailing telluride is petzite, an iron-gray mineral with a black streak verging toward silver gray. In some places it is distributed in irregular bunches, often of considerable size, through a white, dense, granular quartz; it does not form regular crystals. Elsewhere it is disseminated in extremely fine particles through the chalcedonic quartz, to which it gives a dark-gray color. A great deal of gray copper (tetrahedrite) also occurs in the Golden Fleece mine, and when this mineral is in finely divided particles in the ore a distinction between it and the telluride is difficult. It is significant that the tellurides both in the Golden Fleece and in the Gallic-Vulcan mines occur with tetrahedrite and other minerals, such as characterize the prevailing lodes in the region. These lodes are, therefore, in the writers' opinion, to be regarded as local variants of the normal antimonial lead, silver, copper, zinc veins characteristic of this region rather than a separate group of telluride lodes of different age and independent origin.

Quartz.—Quartz (SiO_2) is present in the gangue in all the mines of the Lake City region and is probably also the most abundant vein mineral in the district. It occurs in two sharply contrasted varieties, both of which are present in greater or lesser amount in all of the veins. The first of these is a fine-grained aphanitic variety frequently called jasperoid, which ranges in color from almost black, through all shades of gray, to a dense material having the appearance of porcelain. A good deal of this jasperoid is banded and has apparently resulted from deposition in an open space rather than from replacement. Some of it, however, is characterized by small irregular cavities lined with minute quartz crystals (druses); this form has almost universally resulted from the silicification of the country rock and is a product of replacement rather than of deposition in an open cavity. All of these varieties of jasperoid are extremely tough and dense and their true character can not be determined without the microscope. Finely divided metallic minerals, such as tellurides and tetrahedrite, frequently give a darker color to the fine-grained jasperoid.

The second type of quartz is the ordinary white crystalline variety. This is probably the most common type in most of the veins, in the majority of which it exhibits well-developed comb structure and is the commonest lining of vugs. The white quartz occurs without question in more than one generation, but the bulk of it is younger than any of the metallic minerals, for it cuts through sphalerite, galena, tetrahedrite, pyrite, and even rhodochrosite in places. Veinlets of this white quartz also frequently cement shattered masses of the metallic minerals.

Dolomite and calcite.—Dolomite ($[Ca, Mg])CO_3$) and calcite ($CaCO_3$) occur very rarely as individual species in the Lake City veins. Calcite has been identified by the writers with certainty in only a few places; and both these minerals are of such slight importance in the district as to be negligible. In this respect the Lake City veins differ from those in the Telluride and Silverton districts. It is probable, however, that both calcium and magnesium occur, replacing a portion of the manganese in the rhodocrosite, a mineral which is very common in these veins, and giving it a light pink color. Both calcium and magnesium are present in much of the mixed carbonates of the Ouray district, and it is likely that they occur at Lake City also.

SECONDARY MINERALS.

Atmospheric agencies have acted on the minerals of the Lake City lodes to form oxidation products and secondary sulphide enrichment minerals.

OXIDATION PRODUCTS.

An extensive discussion as to the character of oxidized products formed in the outcropping portions of the Lake City lodes is difficult, if not impossible. Few of the outcrops are prominent, many of them being covered with landslide material and with different forms of rock débris, and such workings as have been driven on them are now generally abandoned and inaccessible. For this reason the list of minerals (p. 34) produced during the processes of oxidation would probably be much increased if a more general examination had been possible.

In general the oxidized zones of the Lake City ore deposits are not deep. Thus in the Moro mine oxidation has penetrated to a depth of approximately 100 feet, and on the Ilma vein, near Lake San Cristobal, to about 200 feet. This lack of depth is due in part to the length of the winter season, which, during a large part of the year, undoubtedly prevents access of water to the veins. However, in spite of the large preponderance of run-off over infiltration, a great deal of water has found its way into the lodes, as is evidenced by the extremely rich character of the oxidized ores and the ores produced by secondary sulphide enrichment. Probably a very considerable part of the eroded portions of the lodes has been carried down in solution into the now existing remnants and has enriched them enough to make profitable mining possible. With less rapid erosion, however, the veins would have been very much richer in their upper portions than is actually the case.

The minerals produced by oxidation are:

1. Minerals such as the soluble sulphates, chalcanthite, and melanterite, which have been taken into solution and afterwards partly crystallized out. These soluble sulphates are rare in the Lake City veins, this probably being due to the continued presence of water in the veins, which has kept the minerals in solution and prevented their crystallization.

2. Minerals formed by the reprecipitation of the dissolved constituents. These are limonite, hematite, basic ferric sulphate, pyrolusite, malachite and azurite in small quantities, anglesite, and cerusite. Cerusite is comparatively uncommon in the Lake City lodes, probably on account of the absence of wall rocks containing carbonate of lime. It is, however, present in some quantity in most of the lodes. Anglesite, especially in its fine-grained form, is a very common oxidation product of the lead veins. It occurs almost exclusively in the upper 100 feet of the Moro vein. Limonite is by far the most abundant of all the minerals produced in this way.

3. Minerals which have been precipitated by the reaction of various sulphates on one another. These are native copper and native silver.

In the Excelsior mine a mass of native copper weighing 150 pounds was found in the oxidized zone, and smaller masses have been found elsewhere, though nowhere in sufficient abundance to be a commercial asset. The copper has probably been produced according to the following equation:

$$CuSO_4 + 2FeSO_4 = Cu + Fe_2(SO_4)_3.$$

Native silver is also a frequent by-product of oxidation processes, having probably been formed in the following manner:

$$Ag_2SO_4 + 2FeSO_4 = 2Ag + Fe_2(SO_4)_3.$$

It has not been found, so far as the writers have been able to determine, below the base of the oxidized ore, but it is distributed in considerable quantity among the oxidized minerals. It occurred in the Gallic-Vulcan, Excelsior, Woodstock, Ute, Hidden Treasure, Ilma, Contention, and Golden Fleece mines, and probably also in many other mines concerning whose oxidized ore no data could be secured. It is usually in the form of wire silver and has probably had considerable importance in rendering profitable the oxidized ores.

Native gold is not common in the oxidized ores of any except the telluride veins, and there only to a minor degree. It is present, however, as will be explained later, in the upper part of the zone of secondary sulphide enrichment.

SECONDARY SULPHIDE ENRICHMENT MINERALS.

Secondary sulphide enrichment has been the most important of all the features that have rendered the Lake City lodes commercially profitable. The comparatively low-grade ores which are found on the levels at depths below the reach of all secondary action and the sharply contrasted and very rich masses of bonanza material in the upper levels are the most striking feature of all of the Lake City lodes. To the failure to recognize this distinction is in large part due the disappointing results of so much of the expensive development in the district. The Vermont tunnel, Red Rover tunnel, and Ilma tunnel (Golden Fleece mine) cost large sums of money and have been extremely disappointing.

The minerals produced by secondary alteration are pyrargyrite and proustite, chalcocite, covellite, galena, bornite, native gold, and probably some argentite.

Pyrargyrite and proustite.—The so-called ruby silver ores occurred in large quantity in all of the producing mines, and to their presence the major part of the silver production of the region is to be attributed. The two ruby silver minerals are the antimonial sulphide, pyrargyrite, Ag_3SbS_3 (59.8 per cent Ag), and proustite, Ag_3AsS_3 (65.5 per cent Ag), the one known as dark ruby silver and the other as light ruby silver. The antimonial variety seems to have been by far the most common in the ores mined, but the arsenical variety undoubtedly occurred, and it is likely that indefinite amounts of arsenic replaced the antimony in much of the ore.

Pyrargyrite and proustite are not at present mined to any extent, as they have long since been exhausted, leaving only the tetrahedrite ores from which they were derived. They occurred either disseminated in crevices and cracks in the sulphide ore, as beautiful crystals in vugs, or as irregular bonanza-like masses segregated along cracks and fissures through which descending solutions penetrated into the unaltered sulphides below. A mass weighing several hundred pounds, taken from the Hidden Treasure mine, was exhibited by Dr. Hoffman in Lake City. This ore was very abundant, intermingled with the telluride ore above the third level in the Golden Fleece and in the adjacent Ilma mine. It occurred, so far as could be learned, in all of the mines at the plane of demarcation between sulphides and oxides and, in generally decreasing quantity, to several hundred feet below this level. Along cracks and fissures it occurred in isolated masses to great depths; for instance, at 1,200 feet in the Golden Fleece and at 1,300 feet in the Ilma. These deep occurrences are, however, uncommon and merely indicate the presence of some easy line of access for downward-moving solutions.

Ruby silver has probably resulted from the solutions of silver and antimony obtained by the decomposition of the tetrahedrite and possibly to some extent also from the argentiferous galena. The chemistry of both the solution and reprecipitation of the antimonial and arsenical sulphur compounds has not yet been worked out in sufficient detail to permit a statement of the probable steps of the process, but the geological facts show that it has occurred. The proofs of the secondary character of the ruby silver are:

1. Its restriction in quantity to the upper levels of the mines.
2. Its invariable occurrence as the latest deposited mineral in the veins, either in cracks or crevices in shattered primary ore or as crystals in cavities.
3. Its occurrence only in isolated bunches in deeper workings, where its origin is probably due to the presence of water channels that permit the downward percolation of water from above.
4. Its complete absence from the great mass of deep-seated ore.

It is clearly secondary in all of a large number of specimens of silver ores from the Southwest.

Secondary chalcocite.—Chalcocite (Cu_2S) in the finely divided sooty form is a common constituent of ores high in chalcopyrite and pyrite; in some mines, the Moro, for instance, it extends as deep as 500 feet. As copper is only a minor ingredient in point of value, this mineral is of more scientific than commercial importance.

Covellite.—The indigo-blue sulphide of copper, covellite (Cu S), does not occur in quantity, but is found in a number of mines as a product of secondary sulphide enrichment coating the surfaces of sphalerite. The sphalerite at first glance appears to be covellite, as it has the luster, color, and general appearance of that mineral. When lightly tapped it falls to pieces, but shows no change in its color or luster, and only when it is broken up quite small does it exhibit the characteristic cleavage and color of sphalerite. It may then be noticed that the films of covellite are not more than one thirty-second to one sixty-fourth of an inch thick and have been deposited all through the sphalerite. As the covellite is less soluble than sphalerite it is clear why the sphalerite has caused the enrichment.

Bornite.—Bornite (Cu_5FeS_4) is rarely seen and then only as thin films on the surface of chalcopyrite which has been exposed to alteration. In massive form it is absent in these mines.

Secondary galena.—Galena (PbS), as a product of secondary enrichment, is uncommon, this being probably due to the fact that oxy-salts of lead are so much more insoluble than the corresponding salts of other metals that their transportation from place to place goes on in only a minor degree. In many mines which have a large content of coarse-grained sphalerite, however, a thin gray film of metallic character has been deposited in the cracks of shattered sphalerite. This material occurred especially in the Pelican mine. The crystalline form of galena could not be detected under the microscope, but from the analogy of this material with some from other localities where the cubes were of sufficient size to recognize, it is probable that the mineral is galena. The sphalerite, which has been thus enriched, may be broken by the hammer without revealing its true character, as it is so penetrated by lead-lined fractures that especial care has to be taken to break a portion of the mass that has not previously been shattered. When this is done the characteristic cleavage and color of sphalerite at once appear.

Gold.—In the Gallic-Vulcan mine crystals of sphalerite were observed coated with leaves of native gold, the latter mineral having apparently been reduced from solution by the zinc sulphide. This occurrence, though of no commercial importance, is worthy of special note as it has been observed by one of the writers in a number of places, notably in the Ibex mine in Leadville, Colo., where metallic gold coating crystals of sphalerite was discovered in large quantity at about the central part of the sulphide-enrichment zone.

PRECIPITATION DUE TO SPHALERITE.

It is interesting to note that zinc blende in the Lake City veins has frequently exerted an extensive effect in re-precipitating downward-moving sulphates in the form of secondary sulphides. It is not believed that the effect produced by this mineral has been sufficiently considered in most papers on secondary enrichment. Its position in the scale of solubilities of the sulphides is of considerable interest in this connection. It was until recently believed that the series representing the order in which the sulphides of one metal would precipitate the sulphides of the other metals was dependent on the relative affinities of the several metals for oxygen and sulphur. It has been recently shown, however, by Wells,[1] and has appeared also in studies by one of the writers that the relative solubilities of the different sulphides was the determining factor in the precipitation of the secondary sulphides. This series of solubilities arranged in decreasing order is as follows: Mercury, silver, copper, antimony, tin, lead, zinc, nickel, cobalt, iron, arsenic, and manganese. From this it appears that, with the exception of pyrite, sphalerite exercises the greatest effect in the production of secondary minerals in the Lake City veins, as nickel, cobalt, and arsenic are there absent. Chalcocite is probably the only secondary mineral that has not been definitely identified in intimate association with zinc blende. The ability of sphalerite to precipitate native gold presumably from a solution of ferric sulphate and chlorine is also interesting. The high silver content of some of the zinc blende encountered in the Monte Queen mine and in some other mines may perhaps also be explained by the presence of native silver precipitated in like manner.

PAY SHOOTS.

All of the Lake City lodes are characterized by the lateral segregation of workable ores within certain more or less restricted portions of the vein. These restricted portions are usually termed pay shoots, as they may be profitably worked, even though the intervening portions are unprofitable. Such localizations of metalliferous minerals may be due (1) to the widening of the fissure between two constricted portions in which neither ore nor gangue minerals have been deposited in any quantity; and (2) to the segregation of metalliferous minerals within the vein.

The first type of shoots is the most common appearing in the Monte Queen, to some extent in the Hidden Treasure, and in many other mines. An interesting feature of the Monte Queen ore body was the different material in the three known shoots. These are vertical and are formed by the swelling of the vein between pinches that leave only a few sheeting planes with some stringers of quartz, etc., by which the vein may be followed from one shoo t to the next. In the Monte Queen there are three of these shoots (figs. 17 and 18, pp. 100 and 101), with stope lengths measured along the strike of the vein of 50, 75, and 200 feet. Nos. 1 and 3 contain chiefly pyrite

[1] Wells, R. C., The fractional precipitation of sulphides: Econ. Geology, vol. 5, No. 1, 1910, pp. 1-14.

and some gray copper ore but do not carry profitable values in silver. No. 2 consists chiefly of a peculiar rosin-like dark-brown sphalerite, mixed with fine steel galena; it averages about 75 ounces of silver to the ton and from 1 to 20 per cent bismuth. A fourth shoot, 950 feet from the mouth of the tunnel, carries a massive, fine-grained aggregate of rhodochrosite.

Within the limits of the No. 2 shoot additional cross shoots occur (fig. 18) which pitch about 15° from the horizontal within the vertical shoot along the course of the vein to the southeast. These interior shoots are lenticular in cross section and have a vertical range of from 3 to 20 feet, with an intervening barren quartz filling of from 3 to 12 feet. (See figs. 17 and 18.)

Some of the shoots have great length, as for instance in the Ute vein. Here the vein from its southwest end nearly to the Hidden Treasure line, a distance of 3,000 feet, was of workable width and the metallic minerals were distributed with remarkable uniformity throughout its whole length, so that nearly this entire distance constituted a shoot. Shoots are reported to have occurred in the Ulay vein, but the workings are now inaccessible.

The second type of pay shoot is that in which a segregation of valuable minerals within a barren gangue occurs. One in the Golden Fleece mine (see fig. 22, p. 107) pitches to the west about 27° and has a length along the vein of about 300 feet at the surface but pinches down to a point below the third tunnel level. Its pitch is about parallel to the trace of the intersection of the Ilma vein, which has a nearly north and south trend and lies mostly just west of the intersection. It is very probable that this intersection has been the determining factor in the origin of the shoot.

In addition to the segregation of the commercially profitable minerals the gangue minerals also show a localized distribution. Thus, in the Monte Queen mine above mentioned, rhodochrosite is rare in the first three shoots but is almost the only mineral present in the fourth shoot, which shows a total width of 5 feet from wall to wall. In the Hidden Treasure, at the northeast end, rhodochrosite and barite greatly preponderate, but the former is almost wholly absent from the Ute portion of the vein and the latter is present in very much smaller amount.

Other details of shoots can not be given, as too few of the mine workings are now accessible for their accurate study.

VALUE OF THE ORES.

The principal product of the Lake City region has been silver, but lead, gold, and copper have to some extent contributed to the value of the ores. Zinc has had practically no part in the production and has been one of the most objectionable components.

If divided according to gold and silver content, the veins of the region might be classed as silver-bearing veins and gold-bearing veins. The tetrahedrite-rhodochrosite group and the quartz-sphalerite-galena group would fall together in the silver-bearing class and the telluride group in the gold-bearing class. The details of ore values can be ascertained only from oral report for many of the mines, because many workings have been abandoned and records could not be secured. Other workings now ship low-grade ores, whereas those mined in early days were of very much higher grade.

In the silver-bearing group the present ores do not average much over $6 to $10 a ton and yield a concentrate worth about $40 a ton. This includes both the high-grade masses of tetrahedrite scattered through the ore and the lower-grade ores that are also a notable feature. In the upper levels of many mines the ore has ranged from $10 a ton up to $3,000 a ton and even more. Eighty-five shipments of ore of about 25 tons each from the upper levels of the Vermont mine averaged 84.53 ounces silver and 27.96 per cent lead, or approximately $44.80 silver and $12.06 lead, a total gross value of $52.86 per ton; 63 tons mined in 1894 yielded $79 per ton; and two shipments in 1895 and 1896 yielded $22.78 and $22.67 a ton, respectively.

The average of the ore mined in the Hidden Treasure mine in 1898–9 was from $7 to $9 per ton, yielding $41.67 per ton of concentrates. The Black Crook in 1884 produced 1,227 tons of ore valued at $124,447, or about $101.39 per ton. All of these rich ores have now been exhausted and the ore must be milled. Probably very little of the ore runs over $5 to $6 per ton.

Of the telluride or gold-bearing veins the Golden Fleece alone has thus far been a producer. Much of the ore it yielded during its early exploitation was of unusually high grade. The following figures, kindly furnished by Mr. George W. Pierce, show the grades of ore mined and milled:

Precious metals, in ounces per ton, from the Golden Fleece mine.

Class.	Gold.	Silver.
1	134.10	3,077
2	6.11	238
3	2.00	53
4	.40	15

The values in all the ores vary so widely that it is not possible to lay down any rule which will not have as many exceptions as it has instances.

SUMMARY OF ORE DEPOSITS.

The lodes of Lake City are fissure veins formed partly through the replacement of shattered and sheeted zones in the country rock and mainly through the filling of open spaces. They average between 500 and 1,000 feet in length, have a similar vertical range, and an average width of 10 inches to 5 feet. They show a wide range of strike and dip at steep angles. They consist of three closely related mineralogic types. The first contains pyrite, galena, sphalerite, and chalcopyrite with subordinate tetrahedrite, in a quartz gangue with some barite and rhodochrosite, and yield silver and lead with subordinate copper and little gold. The second contains galena, sphalerite, and tetrahedrite with subordinate chalcopyrite and pyrite, in a gangue of quartz, barite, and rhodochrosite, and yield chiefly silver and lead. The third contains petzite, tetrahedrite, and minor quantities of other sulphides, in a gangue of fine-grained quartz carrying some hinsdalite in places; this type yields silver and gold in proportion by value of 1:1 and is characterized by high values in both gold and silver.

The first two groups constitute the bulk of the Lake City lodes, as there is but one productive telluride vein. Their ores are low grade where unaffected by surficial alteration, the major portion of the Lake City production coming from ores enriched by secondary sulphide enrichment that has produced chiefly pyrargyrite as a secondary mineral. Oxidized ores are widely distributed. Lodes are, for the most part, separated by considerable areas within which no veins have yet been located, the Capitol City group being the only closely spaced series of veins exploited.

The Lake City lodes comprise the northeastern portion of the mineralized area which includes Rico, Telluride, Silverton, and Ouray. The lodes show great similarity to those of these areas, but were probably formed under a slightly less cover of overlying rocks.

The lodes are later than all of the rocks exposed in the region except the rhyolites and basalts of post-Potosi age and are hence of late Miocene or early Pliocene age. They are believed to have been derived from the vapors emitted from a magma of monzonite whose apophyses as intrusions are scattered through this general region in considerable numbers.

FUTURE OF THE DISTRICT.

The Lake City district possibly contains (p. 44) unworked and undiscovered veins that are similar in mineralogical character to those already developed, and it is on these, combined with the careful mining and milling of the ore from the veins already located, that the future of the district depends. A quantity of low-grade ore is now in sight, associated with which is an enormous amount of zinc blende. This ore may perhaps be treated with profit if the proper kind of concentrating works are erected, and the zinc may also be sold if mechanical means can be successfully used in its separation from the commonly associated minerals.

As the geology of the region has so little apparent connection with the deposition of the ores (p. 46), it is impossible to lay down any rule as to the best places to look for new bodies of mineral. The richest mines are in the andesite of the Picayune volcanic group, but veins found elsewhere in the district have the same general mineralogic content, so that no inference can be drawn from the occurrence in this particular andesite.

The fact that some of the veins can be traced on the surface for considerable distances would seem to point to the general conclusion that the extension of some of the already profitably worked veins may be found by a careful study of the topography and the relation of the veins to it, taking into consideration their dip and strike. On the other hand, as it is rather evident that the district contains mainly only the roots of veins, it would probably be more profitable to look for new discoveries rather than the extension of those already worked, as doubtless the major part of these have been eroded away, leaving only their lower parts.

An interesting feature is the date of the location of the properties which have proved to be successful producers. Without exception the veins which have paid were located in the early days of the mining activity. A great many prospects have been abandoned which show surface croppings just as promising as those of some of the paying properties. Much of the loss that has been sustained in mining in this district has arisen from unjustifiably excessive initial expense. There have been some startling examples of reckless expenditure in the district, which tend to weaken the confidence of the mining public and to destroy any chance of obtaining funds for judicious exploitation.

MINING CONDITIONS.

POWER.

Until recently a majority of the large mines used coal for the generation of power. Lately a few dams have been erected and water power utilized for the generation of electricity for the Hidden Treasure, Moro, and Lellie mines. The first two take water from Henson Creek and the last named from Nellie Creek.

The steep gradient in most of the streams makes it easy to obtain the head necessary for hydroelectric installations in a region of limited water supply. An objection, however, to this means of generating power is the low temperature which prevails during the winter months, at which time the streams throughout the mining district are frozen over. A means of obviating this difficulty is to build a dam at such a place that quite a deep body of water may be obtained, so that when the surface freezes there will be enough water below the crust of ice to furnish the requisite flow. The Hidden Treasure dam is said to be 100 feet in height. The natural conditions of the banks of the stream are admirable for such a dam, and a great supply of water can be maintained.

The officers of the Moro mine state that sufficient water is available to operate the mine during the entire winter, although they think it is doubtful whether they can generate enough power to operate both mine and mill during the months from December to May.

In all probability it would not be possible to obtain water during the winter months from any of the tributaries of Henson Creek, they being frozen solid during the cold season.

The large mines using coal for the generation of power have been the Ute, Ulay, Golden Fleece, Black Crook, and until 1907 the Moro. They have obtained their fuel from Crested Butte, Colo., when possible, and from Somerset, Colo., when the Crested Butte coal was not to be had. Run-of-mine coal sold at the time of writing (January, 1909) at $2.50 to $3 a ton on the cars at the mine, and lump coal at $4.50 a ton in carload lots on the cars at Lake City. Run-of-mine coal is said by some to be very unsatisfactory, but the lump coal is said to be about the most satisfactory in Colorado for steaming purposes.

Gasoline has been used at the Pelican property, but the foundation failed to withstand the vibrations of the engine and further use of the machine was temporarily abandoned. The company intends, however, to place the engine on a firm foundation and resume its use. Gasoline costs 16 cents a gallon in barrel lots on the cars at Pueblo, Colo., and that used by the Chicago Tunnel Site Co. is reported to have averaged 58° to 65° hydrometer test, each shipment, however, showing considerable variation. The Pelican at one time contracted with the Lake City power plant to deliver electricity at the mine. After having erected the line, however, it was found that the company was unable to furnish enough power to run the mine and the scheme was abandoned.

LABOR.

The labor conditions in the Lake City district have been very uniform and the differences between operators and laborers have been slight. When there has been great activity in the region it is understood that labor has been easy to procure. The wages paid are about the same as elsewhere in the San Juan region, ranging from $3 to $5 per day, according to the class of work.

MINING METHODS.

The early work on most of the veins was carried on through drifts driven on the veins. The adit tunnels run were short and easily operated. This was very advantageous during the early stages of exploitation, when the deposits were easily accessible. As depth was gained, however, long crosscut tunnels became necessary and these have almost invariably proved disappointing. The Vermont tunnel, T. C. M. tunnel, Lucky Strike tunnel, Hidden Treasure tunnel, and many others have been started, but so far none have had satisfactory results. Most of them have not gone as far as they had intended to go at the outset, as the character of the vein encountered in depth did not seem to justify the outlay.

No deep vertical shafts are used in the mines, although several properties have developed the veins by inclined shafts following the general dip of the vein.

The ore is transported from the mines by burros, wagons, or wire-rope tramways. The last named are used on the Hidden Treasure, Lellie, and Moro properties to connect the mines with the mills located several hundred feet below the mine openings and from a quarter to three-quarters of a mile distant.

EXTRACTION OF METALS.

From the beginning of active mining operations in the Lake City district there have been smelters and lixiviation plants, which, between them, have treated much of the ore from the Lake City district and some from the Silverton district with more or less success. Because of the historic interest connected with the early operations in the milling and smelting of the ores from this region a few lines will be devoted to that subject.

On October 31, 1877, the Ocean Wave smelter turned out its first lot of base bullion, about 10,000 pounds. The works were equipped with two roasting furnaces and one stack and had a roasting capacity of about 15 tons a day. First charcoal and later coke from Trinidad[1] was used for fuel. Crooke's smelter was also in operation at this time, although details of its equipment are not at hand. It is said to have shipped 300 tons, worth $48,000, during the year 1877.[2] Its slag dump is reported to have been twice reworked with profit. The Van Gieson Lixiviation Works were in successful operation at an early day and are reported to have had an output of $35,000 in silver bars during the fall of the year 1877.[2] The product ranged from 850 to 925 in fineness.[1] Other smelters and other lixiviation plants existed, but a description of them would be of no real value. The only smelter not partly dismantled in the Lake City district in 1909 was the one under construction by the management of the Lake City Mining & Smelting Co. It will have a 100-ton coke furnace, 40-horsepower boiler, 35-horsepower engine, dynamo, 110 volts direct current, blower with 1-pound per square inch pressure, crusher, rolls, grinder, and assay office.

---

[1] Eng. and Min. Jour., vol. 24, 1877, p. 369.
[2] Idem, vol. 25, 1878, p. 62.

The smelters which have received most of the ore from the Lake City district have been the Omaha and Grant smelters of Denver, Colo., the Eilers plant at Pueblo, Colo., and the Ohio and Colorado smelter at Salida, Colo. At times other smelters in the State have received consignments from this area, but the majority has gone to the above-mentioned plants. The freight charges per ton would probably average from $5 to $8 to Salida and Pueblo. The treatment charges plus the cost of mining and shipping will run from $10 to $15 a ton in the Lake City district, so that under existing mining conditions it will hardly pay to work smelting ore running under $20 per ton. With concentration, however, a much lower grade of ore can be mined and milled with profit. The Hidden Treasure, for example, mined and milled $9 ore at a profit. The Golden Fleece milled $12 ore, saving 65 per cent of the value, at a total cost for mining and milling of probably less than $5 per ton.

On the other hand, the processes employed by some of the mills have not proved to be suited to the kind of ore treated, and a large part of the valuable metals was carried into the tailings, several tailing dumps having been worked at a fair profit by various means of lixiviation. The tetrahedrite, most of which is argentiferous, has a specific gravity of only 4.5, and when treated with galena, whose specific gravity is 7.5, it is lost by the same process which saves the lead sulphide. Galena is less brittle than tetrahedrite and does not slime so readily, so that after the ore is crushed and sent through sizers to the tables the galena is easily separated from the rest, and the gray copper, being brittle and of rather low specific gravity compared with the commonly associated minerals, is either left in close association with the gangue or is largely lost in the slimes.

DETAILED DESCRIPTIONS OF MINES.

CAPITOL CITY GROUP.

GALLIC-VULCAN.

The Gallic-Vulcan claims are located on the south side of North Fork of Henson Creek, at an elevation of 10,300+. A good wagon road from Capitol City, about a mile distant, leads directly to the properties. Two patented and six unpatented claims comprise the group.

History and production.—The Vulcan was discovered in 1883 and the Gallic a few years later. The development on the claims has been done almost entirely by Benjamin Guionneau and his relatives, who have at no time employed large shifts or carried on extensive operations. The development is the result of the work of a few men continued over a long period of years.

The total production of the Gallic and Vulcan mines is small. The ore treated has been hand sorted so that no general average of the run-of-mine ore can be obtained. The concentrates produced by a small mill installed on the ground are said to have contained about 0.21 ounce gold, 10 ounces silver, 50 per cent lead, and some zinc and iron.

Development and equipment.—The Vulcan property is developed by an adit tunnel driven about 800 feet to cut the Vulcan vein. A vein, thought to be the Vulcan, was cut about 600 feet in, but not enough work has been done on it to prove its identity. A shallow discovery shaft has been sunk.

The Gallic mine has a crosscut about 820 feet long, which cuts two veins. The first is about 675 feet in and is possibly a continuation of the Vulcan vein; the second is 140 feet farther in. There has been about 1,000 feet of drifting and some stoping on the first vein. On the second vein there is only about 140 feet of work and no stoping.

Most of the ore in the veins has been hand picked. As the quartz is very sugary, the use of powder is unnecessary.

The equipment of the plant consists of a small concentrating building, connected by trestle with the ore house. It contains an old boiler, 40-horsepower engine, table, 1-compartment jig, 2-size screen trommel, one sizer, one elevator, one set of rolls, one crusher, and a small ore bin. From the crusher the ore is raised to a double trommel. The ore which does not go through the first screen is sent to a set of Cornish rolls and thence back to the screens. The fine material goes to the table and the coarse to the jig.

Country rock.—The country rock at the Gallic tunnel is mostly andesite of Silverton volcanic series, which in places contains quartz and hence might be classed as a dacite. At the first vein in the Gallic tunnel the country on one side is andesite and on the other quartz monzonite porphyry. The vein at one place is in the former rock and at others in the latter.

Veins.—The fissure in which the vein occurs shows movement both prior and subsequent to the introduction of the vein material. Surfaces of galena show striations produced by attrition, much movement being noticeable throughout the vein.

The ore shoot is irregular, being wide in some places and almost pinching out entirely in others.

In places the vein contains solid masses of galena several inches wide, but these are not persistent and no definite idea of the actual size of the ore body is obtainable.

The dip of both the Gallic and Vulcan lodes is southwest, the former about 60° to 80° and the latter 50°. The strike of both is N. 7° to 15° W. The trend of the veins is in general rather constant, although numerous marked local changes appear in the workings.

Ores.—The ore comprises galena, sphalerite, pyrite, chalcopyrite, gray copper, hessite, sylvanite, native silver, native tellurium, and native gold, and the gangue, quartz, barite, calcite, some kaolin, and a little fluorite and apatite, the relative abundance being in the order named. The first three ore minerals are by far the most abundant, and quartz probably composes nine-tenths of the gangue.

The quartz is almost universally sugary and full of cavities, many of which are due to the solution of crystals of calcite, barite, or gypsum. Within the cavities in the altered country rock crystals of apatite and fluorite occur with pyrite and sphalerite. Both the ore and gangue minerals show remarkable crystallization. Some tellurides occur in the quartz and a few in thin seams in the country rock near the vein. The native gold has been found in association with tellurium and sphalerite, in places as a partial coating over these minerals.

The original vein was probably a small one of quartz carrying galena, sphalerite, pyrite, and tetrahedrite. Subsequently the vein and the country rock in its immediate vicinity suffered severe deformation and very thorough brecciation. Next the fragments of country rock and vein were cemented by quartz carrying pyrite, galena, sphalerite, chalcopyrite, tellurides, free gold, and silver. This second deposition of quartz and ore minerals did not completely fill the cavities, however, and the resulting vein is spongy and porous throughout.

The wall rock has affected the mineralization slightly, if at all. The first vein in the Gallic tunnel occurs partly in andesite and partly in monzonite porphyry, but neither exerted any noticeable influence on the ore deposit.

The depth of oxidation is probably between 400 and 600 feet. Lead carbonate has been found in the Vulcan crosscut 400 feet below the surface, but none, it is stated, was found in the Gallic workings, which are located about 300 feet below the Vulcan.

EXCELSIOR.

The Excelsior mine is owned by Frank Adams and W. B. Owen, both of Capitol City, Colo., who work the property themselves. It is located on the slope just north of Capitol City, at an elevation of 10,220 feet, and may easily be reached by wagon road. The ore has to be hauled by wagon about 9 miles to Lake City, Colo. The work is all done by hand.

History and production.—The Excelsior was located in 1878 and patented in 1889. The first shipment, made in 1893, consisted of two cars, which brought returns of $65 a ton, averaging 59 ounces of silver and 10 to 12 per cent copper. In 1895 and 1896 shipments were made of about 200 tons, which averaged $65, running about 50 to 59 ounces silver, 11 to 15½ per cent copper, and 5 per cent lead.

From 1896 to the present time (October, 1908) from one to five cars a year have been shipped.

Development.—The workings consist of a crosscut of about 100 feet to the vein and a drift of 500 feet along the vein. Some 200 feet in on the vein the original discovery shaft connecting the mine with the surface is cut. This shaft is no longer in use, the ventilation of the mine being accomplished through the Broker, a contiguous property whose drifts connect with those of the Excelsior on the main level. Above this level considerable work has been done in a stope which extends up about 160 feet and is in general between 90 and 100 feet long. The material taken out of this stope by the Broker and Excelsior companies would amount roughly to 64,000 cubic feet. A smaller stope is located just south of the main large stope above the first level. Below the main drift is a shaft about 106 feet deep, off which short levels have been run, and a body of approximately 75 by 30 by 3 feet has been stoped out. Practically all of the work was carried on in this mine by overhand stoping, the ore and rock being sent down through two mill ways to the main level, and then trammed out in cars, the selected ore going direct to the ore bin and the rest being distributed over the dump, according to its content of gangue and country, into second and third class ore. The country rock broken has been utilized in a large measure to fill the stopes.

Country rock.—The walls are entirely Eureka rhyolite, which shows considerable alteration in the immediate vicinity of the vein and is rather highly impregnated with pyrite. That there has been movement since the beginning of vein filling is evidenced by the inclusion in the vein matter of small pieces of country rock, on either side of which ore minerals occur. The vein as a whole, however, is tightly frozen to the walls.

Veins.—The ore body itself occupies an ordinary fissure and seems to be rather regular in size, averaging perhaps 9 inches in width, and is fairly continuous throughout the workings. In places it is much narrower and elsewhere plays out entirely, leaving only the barren vein. The vein dips 60° to 70° N. 80° E. and strikes in a general direction of N. 10° to 12° W. In the upper workings the dip is more pronounced, approaching the vertical toward the top of the upper stope.

Ores.—The ore minerals are sphalerite in large quantities, a good showing of chalcopyrite (usually occurring with sphalerite) in the upper parts of the mine, some galena, a little gray copper, and much pyrite. According to the owners, 150 pounds of native copper was found 400 feet in on the main level and some native silver was found below the level. Native silver is said to become prominent in the lower workings, where sphalerite predominates. In the upper workings, the top of which is only about 40 or 50 feet below the surface, small quantities of azurite, malachite, and some limonite occur.

Quartz is the only gangue mineral present, the absence of others commonly found in the region being notable. On the main level sulphate of copper is forming on a small scale.

The ore is sorted in the ore house and shipped as first, second, and third class ore direct to the smelters. It is the aim of the operators to ship ore containing at least 39 per cent zinc, so that they will receive returns for this, as well as the lead, silver, and copper.

CZAR.

The Czar mine is now owned by Julius Seymour, of New York, but most of the development on the property was completed prior to his ownership. It is located about half a mile from Capitol City on the eastern slope of Yellowstone Gulch at an elevation of 10,800 feet above sea level and may be reached by wagon road. At present (October, 1908) it is not in operation.

Production and development.—It is reported that two cars shipped to Canon City yielded 22 per cent lead, 26 per cent zinc, and 3 per cent copper, 9 to 14 ounces silver, and $3 gold to the ton. A third car is said to have averaged 39 per cent zinc, 14 per cent lead, 3 or 4 per cent copper, and $3 gold to the ton. In 1904 a small car of galena averaged 42 per cent lead. As the smelters would not pay for both lead and zinc and the sorting of the two ores cost so much it was found that the shipments were unprofitable.

The development work on the property consists of two levels supposed to be drifts on the same vein, and a shaft connecting the upper level with the surface and extending down 120 feet to a junction with an upraise of 100 feet from the lower level. Above the upper level some stoping has been done, over 7,800 cubic feet of rock and vein material having been extracted up to October, 1908; there is also a small underhand stope about 10 feet deep and 30 feet long just beyond the shaft.

Country rock.—The country rock is Eureka rhyolite, which shows little or no alteration a very short distance away from the vein and in all probability has had slight influence on the ore bodies.

Some faulting was observed in the Czar mine and gouge and a small horse of country rock were seen in the breast of the lower level. The whole vein has been pretty well shattered.

Veins.—The ore body is a fissure vein of pyritic lead-zinc ore. The veins comprising it are two in number, one of which is worked in the upper level and the other in the lower. Both are small veins, not over 2 feet wide in any place; the average "ore course" is not more than 3 inches wide, though in places it measures 8 inches and in others narrows to ½ inch. Several small and apparently unimportant veins were seen in the breast of the lower level.

Contrary to the general trend in the district, the vein in the upper level strikes about 6° NE. and dips from 60° to 80° E. The main vein in the lower level has a general northerly strike, with approximately the same dip as that in the upper level. However, its course is quite irregular and no definite direction can be given to it.

Ores.—The ore minerals consist of galena thickly interspersed with sphalerite, the whole frequently cutting through a thin layer of chalcopyrite. Pyrite is a common associate of the quartz, which is the only gangue mineral found. Sphalerite and lead sulphide are about equally abundant and are very intimately associated. All of the ore minerals except the chalcopyrite are fairly well crystallized, but the quartz is almost entirely massive. Probably the chalcopyrite was deposited first, followed by the sphalerite and the galena, the whole being cemented together by quartz and pyrite, which no doubt crystallized contemporaneously.

CAPITOL CITY.

The Capitol City mine is located on the west side of Yellowstone Gulch at an elevation of 10,600 feet and may be reached by wagon road from Capitol City, about a mile distant.

The property has three levels, supposed to be on one vein, which probably extend altogether over 1,700 feet. The actual distance can not be given with accuracy, as the mine was in poor condition at the time of examination, and the uppermost level "caved," so that examination was impossible. Considerable stoping has been done in the mine, especially above the lower level. A mill, called the Capitol City mill, was operated in connection with this mine. It was equipped with a boiler, engine, crusher, rolls, and four jigs. In 1900 this mill was torn down.

The Capitol City ore body is in andesite belonging to the Silverton volcanic series, which shows considerable alteration along the vein. The feldspars are largely changed and the hypersthene crystals are almost entirely gone. Some chloritization is noticeable and a great deal of pyrite is present in the rock. The rock is a lava flow, as are most of the eruptives in the district, and has been more or less disturbed by minor faulting in the vicinity of the vein.

The ore body is similar in many respects to those of the Czar and the Excelsior and may be classed with them. The fissure is clearly defined and fairly constant in direction. It branches or splits up, as shown by the vein in the lower level, which divides after going in about 300 feet. The ore body proper varies in width from

9 to 12 inches and the vein from 6 inches to 2 feet. The strike of the main vein is N. 20° W. and the dip about 80° NE. The western branch of the vein, which splits in the lowest level, is probably the one worked in the middle level. It strikes N. 40° W. and dips about 70° N. 45 E. This branch vein seems more clearly defined and more persistent than that in the lowest level.

The minerals found in the veins are sphalerite, galena, pyrite, and some chalcopyrite in a gangue composed almost entirely of quartz. There are several stages of deposition of the quartz, which is mainly in the massive form, though in many places sugary and in some showing distinct crystal outlines.

YELLOW MEDICINE.

The Yellow Medicine mine is located in Yellowstone Gulch 10,750 feet above sea level and may be reached by wagon road from Capitol City, about 1⅓ miles distant.

Previous to 1896 the production, as given by the mint reports, amounted to approximately $40,000, the greater part of the value coming from silver and about one-eighth from the copper. The production before 1892 and after 1896 is unknown, but during 1896 Crow & Fagan, according to report, shipped 500 tons from the middle level. In 1898 the Yellowstone Mining Co. is said to have shipped concentrates for about three months. In 1907, under bond and lease, O'Brien, Bowles & Bardwell for a time mined and shipped, crude, 15 tons mill dirt a day to the Moro mill. It is reported that the runs through the Moro mill were successful until the price of copper declined. The Moro shipped the results of their treatment as concentrates.

The workings of the mine consist of three levels. Only the upper one was accessible at the time of visit; the middle one was caved at the entrance, and the lower one half filled with ice for 200 feet, beyond which a cave in the roof made further exploration impossible. As a consequence the data contained in this report have been obtained from the upper level, the ore bins, and the dumps. A mill was located just below the lower level and a blacksmith shop on the middle level. The mill was built and started to run in 1897. It contained a boiler, engine, crusher, 3 sets of rolls, revolving screen from each roll, set of revolving screens which went to four Harz jigs, and three buddle tables.

The vein occurs at the contact of a monzonite porphyry intrusion in a pyroxene andesite flow. The latter is very much altered along the contact of the vein and is recognizable only by careful examination.

The ore body occupies a simple fissure. In the upper level it dips 79° N. 70° E. and strikes N. 13° W.

The minerals found in the vein are galena and sphalerite in large quantities (the former predominating), pyrite, and chalcopyrite in a gangue of quartz with a very small amount of barite. In places the galena is very well crystallized in cubes, but much of it is steel galena in large bunches.

LILLY.

The Lilly adit, located at an elevation of about 10,500 feet, was full of water at the time of investigation. It is said that ore has been shipped to the Capitol City mill for concentration, but the results are not definitely known.

CHORD EXTENSION.

The Chord Extension is located at 10,800 feet elevation on the east side of Yellowstone Gulch. Development work consists of a shallow shaft, a crosscut to the vein, and a drift on the vein. Not much stoping has been done. The vein is in monzonite porphyry and is in few places more than 4 inches wide, the ore course varying from ¼ inch to 2 inches. The vein dips 75° N. 75° E. and strikes N. 10° W. According to report all the ore shipped was taken out when the shaft was sunk.

The minerals are tetrahedrite and galena, with some chalcopyrite in a gangue of barite, some quartz, and a little fluorite. The vein shows inclusions of silicified country (?) and some pure hematite. Slickensided surfaces are numerous in the country adjoining the vein, but the vein itself seems very tightly frozen to the walls. There appears to have been first a deposition of chalcopyrite, tetrahedrite, and galena, with a gangue of quartz, followed by movement which disdistorted the vein and a second deposition of minerals in a gangue of barite.

WOODSTOCK.

The Woodstock prospect is located in Yellowstone Gulch at an elevation of 10,100 feet. It is owned by Mr. Whinery, of Lake City, who reports that he shipped 200 pounds in 1895 which brought him $300; almost all of the value came from gold, which is said to have been found native in wire form about 40 feet down the shaft. The developments consist of a shaft 225 feet deep and an adit 75 feet long. These are in pyroxene andesite which shows some little mineralization by pyrite a short distance from the vein. The vein itself contains brecciated fragments of country rock and is made up of a gangue of quartz, with galena, zinc blende, chalcopyrite, and pyrite, in relative amounts corresponding to the order mentioned. The vein is tightly frozen to the wall and varies from 2 to 6 inches, widening as it descends. It dips 70° to 80° N. 70° E. and strikes N. 15° E.

OTTAWA.

The Ottawa prospect is located just above the Woodstock in the same country rock. A thin vein of quartz shows some mineralization and is supposed to have produced 2 or 3 tons of lead ore, which, according to report, were shipped from the vein near the surface in the summer of 1907. The vein dips N. 73° W. and strikes N. 10° E., and has been drifted on for 150 to 200 feet.

TOBY.

The Toby is located a short distance above the Ottawa and has been prospected for 150 to 200 feet by a drift on the vein. The vein, which is mineralized chiefly by oxidized iron ores, strikes N. 17° W. and dips 70° N. 80° E.

SILVER CHORD.

The Silver Chord property is located on the east side of Yellowstone Gulch at an elevation of 10,700 feet in country rock similar to that of the Czar mine. The workings consist of a shaft said to be 80 feet deep, which, at the time of the investigation, was practically full of water. As shown by the inclination of the shaft the vein dips 78° S. 60° E. and strikes S. 10° W. From the appearance of the outcrop the vein is 6 to 18 inches wide and has an ore course from 1 inch to 3 inches wide. The gangue is quartz and the ore minerals are galena, sphalerite, chalcopyrite, and pyrite, an association common to many of the mines and prospects in this gulch. Some ore is reported to have been shipped from this property.

CZARINA.

The Czarina is a small prospect in Yellowstone Gulch just west of the Czar mine. The shaft on the property is said to have been 25 feet deep; rock débris, however, has filled one-third of this. A small open cut on the vein shows it to contain about 2 inches of quartz with no appreciable ore body.

Eastward, across the Czarina property, three small veins outcrop with strikes of N. 6° E., N. 10° E., and N. 6° E.

LUCKY STRIKE TUNNEL.

The Lucky Strike adit tunnel, owned by D. M. Jameson, of Capitol City, Colo., has been run in hopes of cutting in depth the Excelsior, Czar, and other veins worked higher up. The total work done is far in excess of that needed to gain the distance desired. In all over 900 feet of crosscutting has been done to reach a point which could have been reached by 600 feet of straight-ahead work. The crosscut encountered several veins, but they show little or no mineralization.

They had the following dips and strikes in order, from the entrance in: Dip 75° N. 15° W., strike N. 73° E.; dip 83° N. 75° E., strike N. 10° W.; dip 70° S. 10° E., strike N. 80° E.; dip 79° S. 65° W., strike N. 23° W. The last one was practically at the breast of the workings, and the others were approximately one-fourth, two-thirds, and three-fourths of the way in.

The workings pass through Eureka rhyolite for 20 feet, pass a small dike of very much silicified rhyolite, penetrate 120 feet of Eureka rhyolite, cut andesite for a short distance, and pass into monzonite porphyry which prevails for about 500 feet. Then the crosscut swings sharply to the east and the rock becomes characteristic Eureka rhyolite to the breast.

Mr. Jameson has started another crosscut about 250 feet east of the present site to intercept the veins of Yellowstone Gulch properties, and at the present writing (1908) this is in 80 feet.

HENSON CREEK MINES.

The mines of Henson Creek are scattered along Henson Creek from a point about three-fourths of a mile west of Lake City as far as the narrow canyon just west of Capitol City. They are described in order from west to east.

PRIDE OF AMERICA AND BIG CASINO.

A shaft about 60 feet deep has been sunk on a small vein in a gulch which heads south from Henson Creek a little west of the Red Rover mill. It is 500 feet above the creek bed and about 10,000 feet above sea level. Two veins are disclosed in this shaft, a main vein 2 feet wide dipping 60° S. and striking nearly east and a smaller vein striking N. 45° E. Both veins carry galena-freibergite ores, with subordinate sphalerite and but little gangue. The galena appears to be the earlier mineral, as it seems to be broken with areas of silver-bearing tetrahedrite (freibergite) scattered through it. The freibergite is the most valuable mineral and carries very high values in silver; the galena, where unmixed with other minerals, averages only about 12 ounces or less per ton in silver. The assays on this ore showed, according to the owners, from 200 to 412 ounces of silver per ton—a statement which merits belief, as the ore contains a very large quantity of freibergite, which in nearly all of these ores is the silver-bearing mineral.

LELLIE (RED ROVER).

The Lellie mine is located on the north side of Henson Creek, about 1 mile west of the mouth of Pole Creek, on the steep bluff that rises abruptly for 2,500 feet above the bed of the creek. The upper and lower workings (fig. 7) are reached by wagon roads, connecting with the main road in the bed of Henson Creek. The mine is now operated by the Planet Mining & Milling Co.

Production and development.—The Lellie has never been a very heavy producer. From the upper levels it has produced some ore, which, according to the management, was high in grade. The mint reports[1] show that in 1899 it was one of the important producers, yielding from the upper stope four cars of ore per week, which netted about $800 per car; but as the amount of stoping is small and it receives no specific mention in previous or later years, its active production seems to have been confined to that year. At the time of examination (1904) and since that time it has been

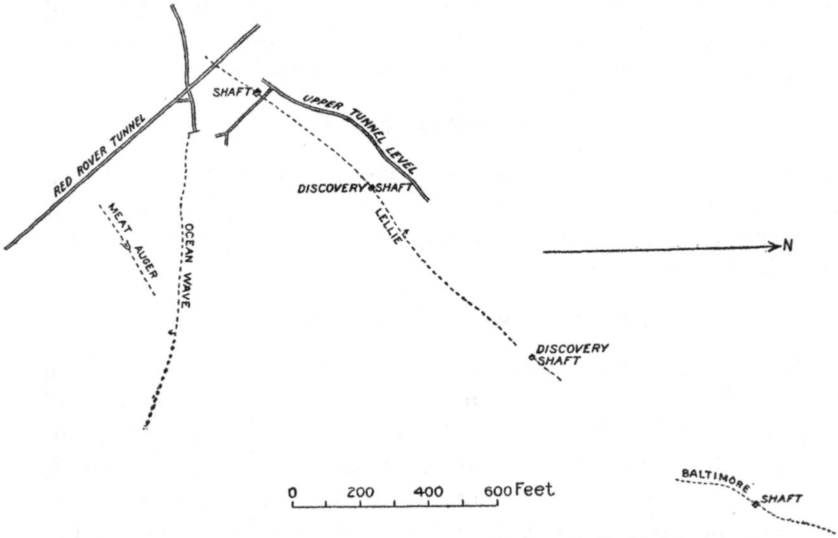

FIGURE 7.—Workings and outcrops of veins of Ocean Wave and Lellie (Red Rover) mines.

only intermittently worked. It was first opened by the upper crosscut in 1897.

The mine is fully equipped with accommodations for miners, shops, and a 50-ton concentrating mill, situated next to the main road in the bed of Henson Creek and operated by electricity. Water power is furnished to the mill by a 2-foot pipe line 3,700 feet in length, with a head of nearly 400 feet.[2] A wire-rope tramway connects the mine with the mill.

The vein is opened by two crosscut tunnels. The upper, which was the first driven, is a short N. 43° W. adit 200 feet long at right angles to the strike of the vein; the lower tunnel, which is 500 feet below at an elevation of 9,650 feet above sea level and about 260 feet above the bed of Henson Creek, is 1,140 feet long and taps the main vein in

[1] Report of the Director of the Mint for 1899, p. 112.

[2] Rept. State Com. Min. for 1901–2, p. 92.

its far end. Drifts have been driven northeast on both levels for about 700 feet. These two levels are connected by a raise 500 feet high (vertical), from which three intermediate levels, separated by vertical distances of 120 feet, have been begun. Little drifting, however, has been done on them, as the vein was uniformly low grade and narrow. Some crosscutting and exploration on two parallel veins lying south of the main vein have apparently been unsatisfactory.

Here, as elsewhere in the district, the stoping near the surface and the high grade of the ore there encountered have led to the driving of a long (1,140 feet) crosscut for a vein which, when cut, failed to meet expectations.

Country rock.—The country rocks of the veins belong to the Picayune volcanic group, which consists, as stated by Mr. Cross, of layers of volcanic flow breccias, both andesitic and rhyolitic, whose rude stratification is well shown in the Red Rover crosscut tunnel. These beds vary in dip from a comparatively slight northerly inclination to one of 45°. For the most part they are less steep than the veins, but they seem to have influenced the formation of these to some extent. The veins steepen up in places and run parallel to the layers of breccia, giving the impression that they are replacements along especially favorable beds. At the far north end of the tunnel the sheeting planes of the vein coincide with the stratification of the Picayune volcanic group (which dips 45° N.), making recognition of the true character of the vein difficult.

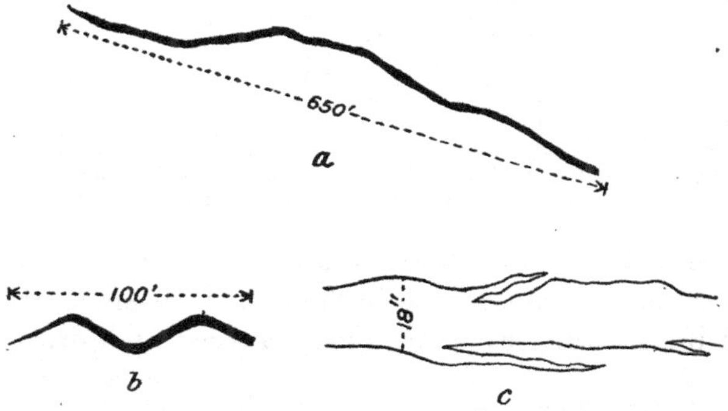

FIGURE 8.—Irregularities in the Lellie fissure along the strike. *a*, Plan of vein in long drift; *b*, plan of vein in short drift; *c*, plan of small portion of vein, showing irregular branches.

As replacement has for the most part been subordinate to the filling of openings, the wall rock seems to have exerted but little effect on the deposition of the ore, but until the different rock masses that make up the Picayune volcanic group have been carefully platted and mapped, the actual relations can not be stated with certainty.

Veins.—The Lellie veins are little if at all disturbed by faulting. The ore deposits mined are typical fissure veins. Replacement here, as elsewhere, has been subordinate, but is more marked than in most of the veins of the region. In the upper workings, where for 400 feet the vein averaged from 1 foot to 4 feet in width, it has fairly well-defined walls frozen tightly to the filling material, with little or no clay selvage present. In the lower workings the vein is much more irregular, being generally only 8 to 14 inches wide or being broken up into a large number of parallel and intersecting stringers separated by irregular sheets of country rock. The vein dips about 60° NW., but in the bottom level is parallel to the stratification of the flow breccia, which dips 45° N. In strike also the vein is extremely irregular (fig. 8).

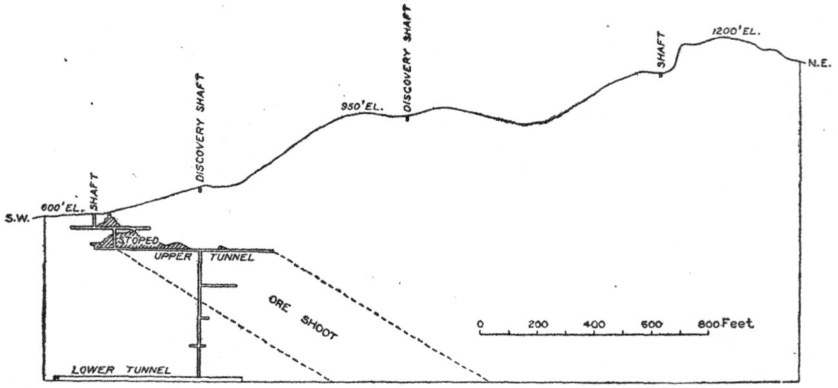

FIGURE 9.—Longitudinal section of the Lellie vein, showing position of rich upper stopes and inclination of the ore shoot. Elevations given are vertical distances above the lower tunnel level.

In detail the vein shows many angular inclusions of country rock, and is for the most part broken up into small irregular stringers, with intervening layers of highly altered andesite. It is difficult to compare the vein in the upper workings with that below on account of the inaccessible condition of the old abandoned upper portions of the mine; but it is probable that in this mine, as in most others in the district, the wide and fairly workable veins die out in depth into narrow and stringer-like roots that rarely pay for the added expense of deeper exploration and operation.

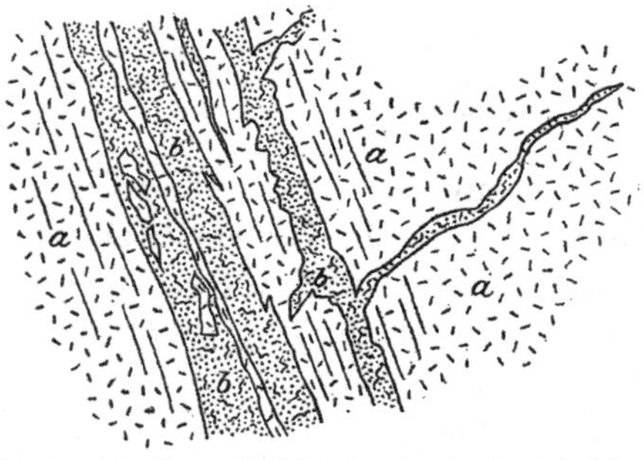

FIGURE 10.—Face of Lellie vein, showing irregularities of vein structure. *a*, Country rock with planes of sheeting; *b*, vein filling.

Ores.—The vein fillings consist of quartz, rhodochrosite, barite, pyrite, galena, sphalerite, and tetrahedrite, with some chalcopyrite in the workings below the first level. The galena, when abundant and massive, carries as high as 20 ounces of silver per ton, though it often runs below that; but the great bulk of the values lies in the argentiferous tetrahedrite, which is readily distinguishable by its greasy appearance. In places it is mixed with the galena and in places scattered separately through the gangue. Pronounced banding is to be seen here and there.

Some of the ore that carries chalcopyrite runs well in gold, but otherwise the gold content is not noteworthy. The high-grade ore from the upper levels consisted of an intricate mixture of argentiferous tetrahedrite and galena. The nature of the ore in these levels is not known, but it is reported to have carried high values in silver and, locally, in gold.

Some of the ore, consisting chiefly of freibergite, averages per ton: Silver 200 to 1,000 ounces, value $500;[1] copper 12 per cent, value $33; gold 1.5 ounces, value $30; a total value of $563.

The longitudinal section of the vein (fig. 9) shows that the ore occurred in shoots which pitched 30° NE. along the strike of the vein. These were separated by barren areas in which the gangue of quartz predominated and the metallic minerals were relatively less abundant. The shoots had a stope length of about 250 feet. The rich ore all occurred in stopes which were near the surface and well within the zone of secondary sulphide enrichment. Figure 10 shows some irregularities of the vein structure.

VERMONT.

The Vermont mine is located on the west side of El Paso Creek about half a mile north of Henson Creek. It is one of four claims, the Scotia, Vermont, Ocean Wave, and Wave-of-the-Ocean, that lie end to end in a S. 75° W. direction on what is in all probability the same vein (fig. 11).

Development and production.—The openings are at the base of a high precipitous bluff that forms the west wall of the creek, towering upward for 500 feet. A shaft 175 feet deep is reported but was not seen. The workings are about 10,000 feet above sea and about 500 feet above Henson Creek. A rough wagon road leads from the mine down to Henson Creek, where it joins the main road to Lake City. The mine is opened by a S. 65° W. tunnel into the bluff. The upper workings have been abandoned since 1899 and only this drift is now accessible, but the mine has in the past produced a considerable tonnage of ore. The high values found in the upper workings induced the management to attempt a long adit tunnel from Henson Creek to cut the main vein. This was driven in for 1,500 feet in 1899 at a heavy expense and was then abandoned.

The Vermont is one of the older mines in the Lake City region. Work on it has been abandoned since 1906. The production has been about as follows:

---

[1] Values figured on price of silver during years when mined and on tonnage shipped.

Production of Vermont mine.

[From mint reports.]

1884	$4,965.00
1887	40,436.00
1888	62,648.00
1892	6,993.00
1895	[1] 182.24
1896	[1] 119.60
	$115,343.84

The mine records for the same years show a production of $115,043.84.

Ores.—No data are available as to the mineralogical character of the ore from the Vermont mine, but as it is of the same general vein series as the closely associated veins of the Lellie, Wave-of-the-Ocean, and Ocean Wave, it undoubtedly consisted chiefly of argentiferous tetrahedrite and galena, with smaller quantities of sphalerite, chalcopyrite, and pyrite. The main output of the mine has been silver and lead. The value of the ore from the upper workings in 85 shipments of 25 tons each ranged from 33.9 to 253.10 ounces of silver, and 7.5 to 59 per cent lead. The average silver and lead content for these 85 shipments was 84.53 ounces of silver and 27.96 per cent lead, giving an average value, computed on the present market prices, of $44.80 silver and $12.06 lead, or a total average gross value of $56.86 per ton of ore mined. In 1894, 63 tons mined yielded $4,965, an average of about $79 per ton. Two later shipments of ore of 16,078 pounds and 10,504 pounds, in 1895 and 1896, respectively, yielded an average total value of $22.67 and $22.78 per ton each.

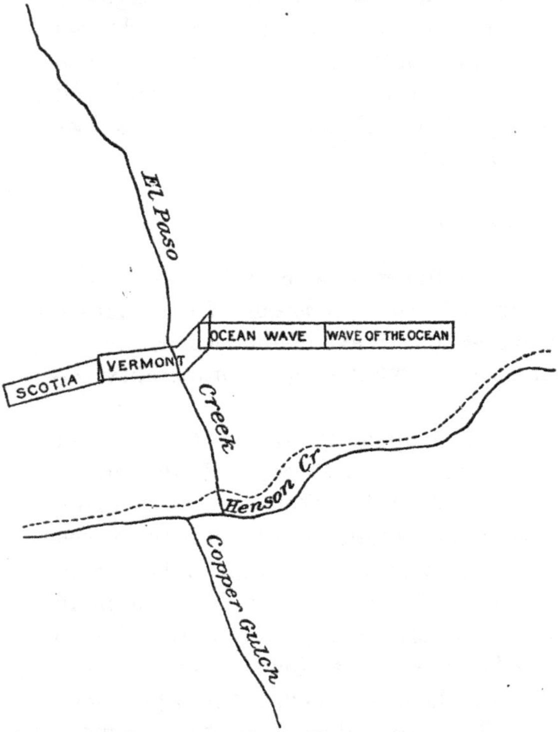

FIGURE 11.—Sketch of the Scotia, Vermont, Ocean Wave, and Wave-of-the-Ocean group of claims.

[1] Figures furnished by the company.

UTE AND ULAY AND HIDDEN TREASURE.

The most productive and also the most extensive workings in the Lake City region are those of the Ute and Ulay and the Hidden Treasure mines. The two, though operated by separate companies, are to be regarded as a single unit geologically, inasmuch as they are on the same vein or group of veins, and their workings interpenetrate one another. Plate V shows a plan of the underground workings and outcrops on the Ute and Ulay and the Hidden Treasure ground.

DEVELOPMENT.

UTE AND ULAY MINE.

The workings of the Ute and Ulay mine are located in and adjacent to Henson Creek, at the eastern extremity of the town of Henson, a settlement which has grown up in consequence of the operation of these mines.

Four veins have been extensively worked—the Ute vein, the Ulay vein, the Annie vein, and an unnamed vein cut in the new shaft. The earlier workings of the mine were on the Ulay vein, which crosses the bed of Henson Creek. Two short drifts above the creek on the north side enter directly on the outcrop of the vein, but these are not extensive, and but little mining has been done on them. The main working opening is an inclined shaft dipping northwest. From it nine levels have been run. The first three, called third, fourth, and fifth levels, are not extensive. The sixth is more extensive, but the seventh, eighth, ninth, tenth, and eleventh are the main working levels of the mine and from them the large production of the vein has been stoped.

All of these workings are in bad repair and the mine is full of water, so that no description of the Ulay vein can be given other than that furnished by those who have worked in it and operated it.

The workings of the Ute vein interpenetrate those of the Hidden Treasure to the northeast. The vein (Pl. V) lies about 500 feet northwest of the Ulay and is roughly parallel to it. It is opened only on the hill slope northeast of the Ulay shaft and has not been followed southwestward across the creek. It is opened by two crosscut tunnels to the first and second levels. The fourth level is opened by an extremely crooked adit running in from the hillside northeast of the Ulay shaft. The lowest level is entered by a long crosscut tunnel from a point 70 feet northeast of the Ulay shaft. From the most southwesterly point in the mine the workings extend to the Hidden Treasure line. Throughout the greater part of this distance the ground is stoped, only a few pillars of lean ore being left to support the roof. No stope maps or very reliable information could be secured about these workings and statements as to the ores can be based only on inferences drawn from the small amounts of ore still remaining.

At the northeastern end of the mine the Ute vein passes into the Hidden Treasure ground, the third level of the Ute being continuous with the fifth level of the Hidden Treasure.

HIDDEN TREASURE MINE.

The Hidden Treasure property includes the five claims, Hidden Treasure, Invincible, Protector, Don Quixote, and Crystal Crown. It

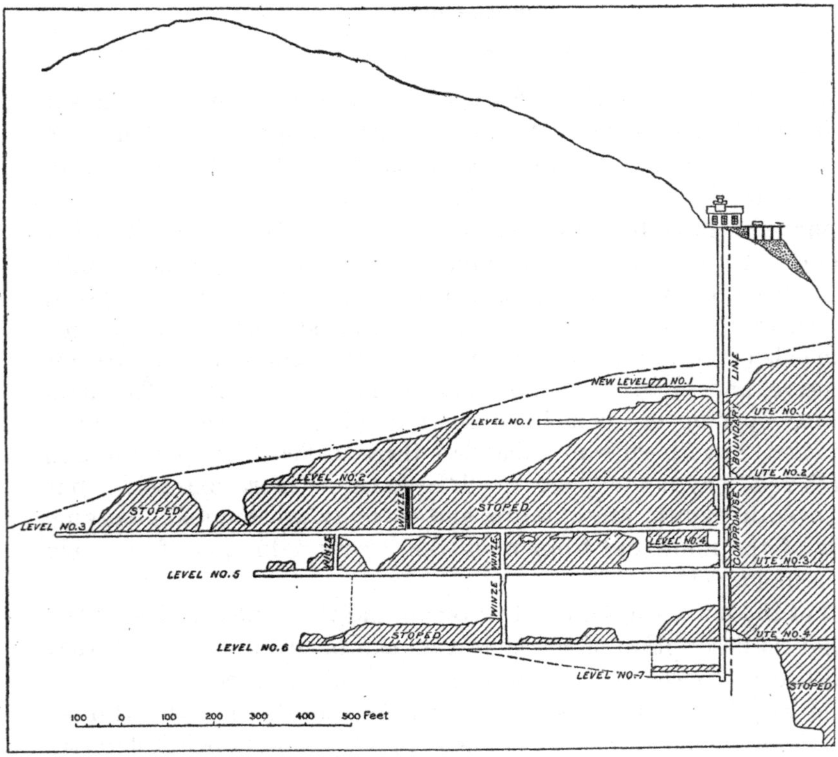

FIGURE 12.—Longitudinal section of the Hidden Treasure vein, showing the stopes and the pitch of the ore shoot.

is equipped with a 100-ton concentrating mill erected in 1898, a water-driven compressor, a 50-horsepower electric hoist, and a Bleichert wire-rope tram, with twenty-five 450 to 500 pound buckets, having a capacity of about 12 tons per hour, connecting the shaft with the mill.

The shaft was sunk on the vein for 265 feet before it encountered ore which paid expenses. The first ore was taken from the mine in June, 1897, and from that time until 1908 it has been worked almost continuously. During 1897 ore was shipped to the smelters, as the concentration plant had not then been erected, but since 1897 the bulk of the ore has been milled and concentrates averaging about $40 a ton have been shipped. Eight levels have been run from the shaft. (See fig. 12.) The vein has been stoped almost continuously from the fifth level to the surface.

The Ute and Ulay mine has been among the largest producers of silver and lead in Colorado, being reported to have yielded between $10,000,000 and $12,000,000 gross. This production takes no account of the poorly adapted milling plant in which much of the earlier ore was treated and by which it is estimated that a great deal of money was lost. The Hidden Treasure mine has produced upward of $700,000.

COUNTRY ROCKS.

The country rock in which the ore bodies occur is the Picayune volcanic group, which carries the largest number of valuable deposits in the Lake City region. This rock shows a great variety of texture and color, so that its identification in the several localities where the best-known mines occur would be impossible without painstaking geological labor in tracing it from one locality to another. Both the Ute-Hidden Treasure vein and the Ulay vein lie wholly within this rock, although at the northeast end the Hidden Treasure vein closely approaches the overlying Eureka rhyolite. In the deeper portions of the workings and in the mouth of the long adit tunnel where the rock is unaltered by surface agencies it has a dense grayish appearance, almost glassy in places. It is characterized by broken brecciated fragments, which are of the same nature as the matrix which incloses them and which were obviously formed during the flow of the rock as it cooled. The congealing crust was broken during the flow and the fragments submerged in the still molten portions, giving rise to a breccia in which no distinction can be made between matrix and fragments. The fragments are of all sizes, varying from almost imperceptible irregular flakes to those measuring many inches. The feldspars are small, though in places they perceptibly mottle the fresh rock. In the vicinity of the Ute vein the rock (see p. 44) has been profoundly altered, passing from gray to almost black, and the black giving way again, in the immediate contact with the vein filling, to a light green, which is chiefly due to the silicification which has accompanied the ore deposition.

Small fragments of country rock, when included in the ore, are generally entirely altered and no black cores are perceptible, but large fragments, which alteration has been unable completely to penetrate, consist of a light-greenish shell over a dark blackish core. The rock is brittle and rather easily ruptured, so that regular veins have formed in it (as in the breccia of the San Juan formation in the

Ouray district), whereas in other tougher types of rock, such as the coarse even-grained andesites in the vicinity of Capitol City, the veins are more discontinuous and constricted. Another consequence of the weakness of the rock is the very large number of horses, or fragments of country rock, found in the ore—perhaps as striking a feature of these Lake City veins as any observed.

ORE BODIES.

VEINS.

The ore bodies of the Ute and Ulay and the Hidden Treasure mines are fissure veins. They are distinctly the result of the filling of open fissures and represent only in minor degree a replacement of the wall rock. Silicification, as will later be shown, in many places affects the walls and included fragments of country rock, but metalliferous minerals are confined to the filling material. Four well-defined veins have been either partly or extensively explored on the property of the Hidden Treasure and Ute and Ulay mines. (See Pl. V.) In the order of their importance they are the Ute, the Ulay, the Annie, and an unnamed vein encountered in the new shaft. A small and yet undeveloped vein known as the California should also be mentioned. An adit tunnel 500 feet in length has been driven to intersect it. The Annie, the Ulay, and the new vein were, at the time of the visit, entirely inaccessible to examination, so that little can be learned concerning them except from the mine maps.

UTE VEIN.

Strike and dip.—The Ute and the Ulay are the main veins exposed on the property. The Ute has been by far the most productive and important of any vein in the district. It has been traced on the surface for a distance of 2,700 feet along the outcrop, from a point northeast of the Hidden Treasure shaft southwest to a tram house just above the mouth of the third level tunnel of the Ute, 1,200 feet north of Henson Creek. Beyond these two limits it has not yet been traced in either direction.

The vein, as explored in the workings, describes an arc of large radius, with the concave side toward the northwest (in the direction of dip). At the southwest end the strike is N. 46° E., in the center it is N. 37° E., and at the northeast end beyond the Hidden Treasure shaft it is N. 19° E. The concavity of the outcrop is increased by a shallowing of the dip in the center of the Ute workings. This is brought out on Pl. V, in which the different levels, which may be regarded as contours, may be seen to close up at both ends and flare apart toward the center of the vein's course.

In detail the strike is irregular, as the vein shows many slight deflections in strike as well as in dip, in this respect resembling the majority of fissure veins. These deflections are rarely large; the greatest is that which shows on the third and fourth levels of the Ute mine, 325 feet north of the adit tunnel of the second level, where the vein diverges 45°, continuing so for 45 feet, and then returns to its original direction. In outcrop the course of the vein is much more sinuous and diverges to the north from the true strike on account of the steep and irregular nature of the topography which it intersects.

Sharp bends occur in the dip also, a fact amply demonstrated in the Hidden Treasure shaft, where the skip now descends vertically, then moves almost horizontally, and then again goes down at a very steep angle. The average dip is between 56° and 60° NW.

Movement.—No criteria exist by which to determine definitely how much differential movement has occurred between the walls of the vein. The variation in width where pinches and swells follow one another in the direction of dip and strike indicate that there has been movement, but that it has not been excessive. The vein is characterized by considerable brecciated country rock which is included in the vein filling, but the brecciation has taken place on a relatively small scale. The fragments of wall rock are generally separated from one another by considerable distances and no evidences of excessive attrition are present. Faulting has not seriously affected the vein. A few cross veins are noticeable, but they seem to be of the same age as the main fissure. This regular and uninterrupted character has done much to facilitate exploration and to make expensive dead work unnecessary.

Width.—The vein averages about 4 feet in width, but pinches and swells after the manner of most fissure veins. The maximum width from wall to wall is about 20 feet, the minimum a mere fracture or series of fractures with practically no width. The wide places occur (1). from the splitting of the vein; and (2) from the movement of the two walls past one another, so that two concave surfaces are brought into juxtaposition. Such wide places run down to mere fissures at both ends, where convex surfaces are in contact; narrow, barren faces may be observed in the southwest end of the Ute workings and in the northeast end of the Hidden Treasure workings.

Through the Ute mine the vein is well defined between two simple walls, but in the Hidden Treasure mine it splits into a number of stringers. At the most important of such splits occurred one of the widest and most valuable bodies of ore. This split is most noticeable on the fifth level, 45 feet northeast of the shaft. One branch of the vein is 10 feet in width, and the other is 4 feet in width. The main vein for 40 feet southwest of the split is 20 feet wide. The junction between these two branches pitches northeast in the plane of the vein, so that on the fourth level it is close to the shaft.

Another irregularity occurs at the southwest end of the Ute workings on the fifth level, where the vein is split into a number of parallel stringers of small size. In a short crosscut northwest from the adit tunnel these stringers are very clearly shown separated from the main vein by spaces of barren but silicified rock; where cut they are 40 feet southeast of the main vein, which here narrows to a maximum width of 2½ feet.

PARAGENESIS.

A number of well-defined steps in the formation of the Ute-Hidden Treasure ore bodies are evident. The first seems to have been the rupture of the country rock with a little but not marked brecciation and but little or no separation of the walls from one another. This was followed by the entrance of silicious mineralizing waters, intensely altering the country rock into hard, greenish silicified material for a short distance from each individual fissure. Later movement separated the walls and moved them slightly past one another, producing the pinches and swells and causing some brecciation of the wall rock.

The first period of vein filling then occurred depositing (1) quartz; (2) rhodochrosite, tetrahedrite, and galena; and (3) quartz. Later movement shattered this vein material and deposited (1) quartz; (2) barite with subordinate galena; and (3) quartz.

ORES.

Character.—The ore in the Ulay vein can not now be described exactly, but it was composed, as nearly as can be learned, of argentiferous galena, with some tetrahedrite, sphalerite, and pyrite, and the usual enrichments of ruby silver. From the upper levels considerable native silver is reported to have been taken. In the lowest or eleventh level the sphalerite increased in quantity and the silver and lead values fell off to such a degree that the mine was abandoned and has not been worked for many years. Some idea of the prevailing character of the ore may be learned from the production statistics for the years 1887 and 1888.

Production of the Ulay vein for 1887–88.

	Gold.	Silver.	Lead.
1887	$1,645	$39,498	$19,242
1888	2,500	84,038	52,800

The absence of copper and the large value of the lead indicate that the product was chiefly argentiferous galena and that there was less tetrahedrite in the ore than in most of the mines of this region.

The ore in the Ute vein was a mixture of galena, pyrite, and subordinate tetrahedrite in a gangue composed mostly of quartz, with subordinate barite and a very small quantity of manganese-bearing minerals. In the pillars left in the mine banded ore containing chiefly galena and sphalerite was observed. In the breasts of No. 5 level about 180 feet southeast of the Hidden Treasure shaft the vein is 4 feet wide and consists mainly of quartz and barite. Many fragments of country rock, generally altered to a very light greenish color, may be seen in the vein.

The sphalerite, galena, and barite of the Ute vein appear to belong to the earlier period of mineralization; in all cases where they could be clearly observed they are older than the gangue of white quartz in which they are embedded. This can be seen from the manner in which a great part of the galena is fractured and broken, and the interstices filled with white quartz, much of it with well-developed comb structure perpendicular to the galena fragments. (See fig. 13.)

Value.—Small bunches and bonanzas of ruby silver in the upper stopes of the Hidden Treasure carried high values in silver, but the general average of the ore extracted from the mine was from $7 to $9 per ton. Milling operations during 1898 and 1899 showed that 1 ton of concentrates, worth $41.67, was obtained from 4.61 tons of crude ore. The cost of operation at the Hidden Treasure, including freight and smelter charges, was about $4.12.

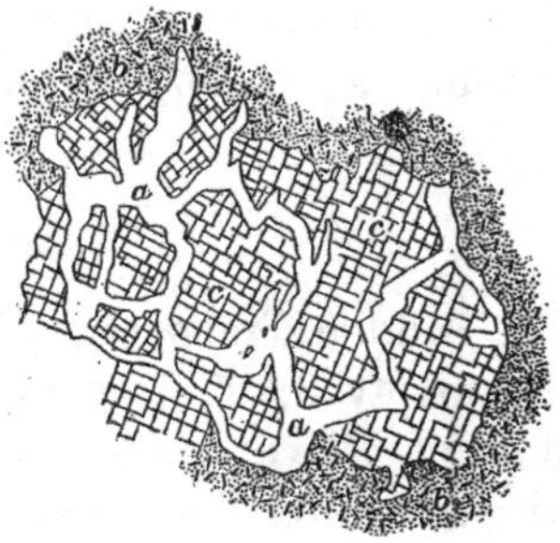

FIGURE 13.—Galena shattered and infiltrated with quartz, Ute vein. *a*, Quartz; *b*, sphalerite; *c*, galena.

Distribution of values.—The distribution of the various minerals in the Ute-Hidden Treasure vein is not uniform. From the southwestern extremity of the Ute vein to a point well within the lines of the Hidden Treasure property, where the crescent-shaped bend in the vein causes it to trend more nearly north, the ore consisted of a banded aggregate of argentiferous galena, sphalerite, quartz, and barite. Throughout this entire distance tetrahedrite is present in small quantities, as is also barite, but rhodochrosite is practically absent. If the workings on the vein had been carried no farther than the Hidden Treasure line, it would have seemed a remarkably clear type of a quartz vein carrying argentiferous galena and sphalerite. Beyond the Hidden Treasure line, however, the barite increases notably in amount and tetrahedrite and rhodochrosite begin to appear in the ore; and finally nearly the whole vein consists of rhodochrosite, with large quantities of tetrahedrite and barite and small quantities of galena. Quartz is present in subordinate amounts. Throughout the long Ute portion of the vein, 3,000 feet in all, the gangue and ore minerals are mingled together with remarkable uniformity, so that for nearly the whole of this length the vein is stoped. No definite pay shoot can be detected, though low-grade areas of small dimensions were encountered here and there and the ore in them was left standing in the mine. In the Hidden Treasure portion of the vein the upper edge of the profitable ore pitches about 15° N., but this pitch is due more to a decrease in the size of the fissure than to the failure of metalliferous mineral contents. The presence also of the branch vein described on page 91 seems to have determined the upper edge of this profitable area. Much difference of opinion among those who operated these properties exists as to the pitch of the shoots, but this arises from the very large area of profitable ore and the consequent difficulty in determining any well-defined shoots. If rhodochrosite and tetrahedrite be regarded as localized from the other minerals in the mine, they are clearly restricted to the north end, but so far as silver values are concerned there seems to be no regularity about the localization.

The ore in the Ulay vein occurred in distinct shoots, two of which were found south and one north of the shaft.

EQUIPMENT.

HIDDEN TREASURE MINE.

The Hidden Treasure mill is connected with the shaft by a Bleichert tramway 3,800 feet in length, with a capacity of 100 tons in 24 hours. Power is obtained by a 150-horsepower Pelton wheel supplied with water from a dam under a head of 118 feet by a wooden pipe line 28 inches in diameter and 1,000 feet in length. The ore goes first through a 9 by 15 inch crusher and thence to rolls, one set 16 by 24 inches, the other 12 by 20 inches, for finer crushing. From here it is sent to a set of trommels, where it is sized, the oversize from the first three trommels being returned to the rolls. The two sizes from the last two trommels pass to two sets of jigs. The fines are sent to two hydraulic sizers and thence to four Wilfley tables, which yield a fair saving. A little of the rhodochrosite is saved from the tables, as it contains high values in silver from finely-divided tetrahedrite. The concentrates are not dried. The concentration is 4.6 to 1. Compressed air is supplied to the mine by a 13½-inch Leyner duplex compressor of 16-inch stroke, which is operated by another 150-horsepower 54-inch Pelton wheel. The hoist is operated by electricity furnished by a 45-kilowatt dynamo, operated by steam.

UTE AND ULAY MINE.

The ore from the Ute and Ulay mine was taken to the mill in Henson Creek by a wire-rope tram. The mill has a capacity of 90 to 100 tons a day. The ore goes first to a rock breaker (Blake, 9 by

15 inches), and then to three sets of rolls (Allis-Chalmers, 16 by 30 inches), then through four successive trommels, 36 inches in diameter and 7 feet long, which size the crushed ore to 8, 6, 4, and 2½ millimeters. The coarse which passes through the trommels goes to the jigs, a double-compartment jig for each trommel. The fines which escape from the last trommel pass into two hydraulic sizers, the coarse being sent to jigs and the fines going into a third sizer. The coarse from this last sizer goes to a jig and the fines run to the buddles, two of which are plain, 16 feet in diameter, and four double-deck, 24 feet in diameter. The tailings pass into settling tanks where the slime is arrested. (See fig. 14.) The Ute ore became much higher in sphalerite on the fifth level, so that the mine no longer paid expenses of operation. In this respect it resembles most of the other veins of the region.

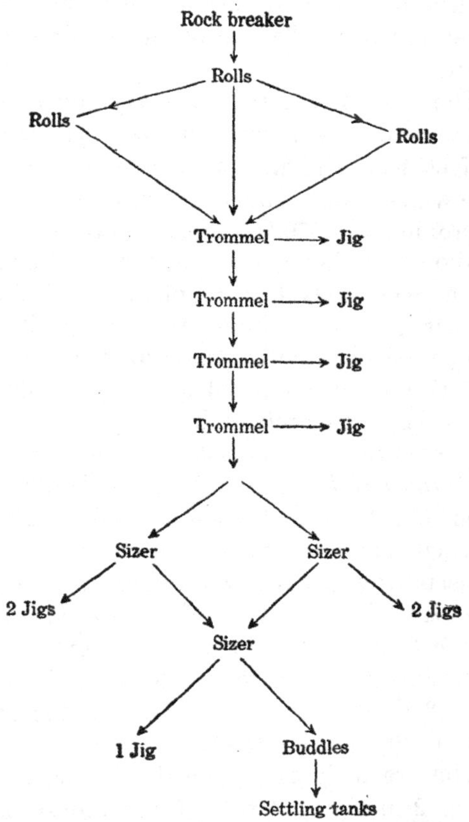

FIGURE 14.—Diagram of treatment at the Ute and Ulay mill. After Rickard.

The following statement by Rickard[1] will indicate the manner in which the concentrates are handled:

The concentrates are dried and mixed by passing through a heated revolving cylinder. About 1½ per cent of moisture is left in the concentrates in order to lessen the leakage arising from the bad flooring of the railroad cars, which would be a greater source of loss if the concentrates were dry enough to run readily. The concentrates contain 58 to 61 per cent lead, 13 to 15 ounces of silver, and 0.05 to 0.06 ounce of gold per ton. They represent about 16 per cent in weight of the original ore and an extraction of about 80 per cent of the lead and 65 per cent of the silver.

PELICAN.

The Pelican mine is situated in a small dry gulch on the north side of Henson Creek about 1¼ miles due west of Lake City, at an elevation of 9,630 feet above sea level and 875 feet above the bottom of Henson Creek. The mine is connected with the main wagon road in Henson

[1] Rickard, T. A., Across the San Juan Mountains: Eng. and Min. Jour., vol. 76, 1903, pp. 461-462.

Creek by a rough second-class road about half a mile long, down the north side of the canyon. Power has been furnished by a gasoline engine.

Production and development.—The first mention of the Pelican in the mint reports occurred in 1891, during which year it is credited with a production of $581 in silver. The following year it produced $1,503, all silver except $20 in gold.

It is reported that two shipments of about 10 tons each showed returns of 0.69 and 0.63 ounce of gold and 165 and 129 ounces of silver per ton. To one of these shipments 2.07 per cent of lead is credited.

The workings of this property consist of two shafts, two levels, a winze, and several small stopes. The shafts are respectively 75 and 38 feet deep, the first one connecting the first level with the surface. The winze is only 35 feet down from the first level. The first level is approximately 450 feet long, and the second (200 feet below the first), is about 900 feet long. All the workings except a 400-foot crosscut on the second level are on one or another of the veins which form the fracture system making up this deposit. The amount of stoping is comparatively small. One block, approximately 28 by 38 by 4 feet, had (1908) been worked at the shaft on the upper level. Another block, approximately 56 by 50 by 2½ feet, had (at the same date) been extracted near the breast of the first level.

Country rock.—The vein is in Eureka rhyolite, the latter having been slightly altered along its contact with the mineralizing agencies. Marked disturbances were noted in the country rock between the crosscut and the vein worked in the second level.

Veins.—In the stope above the lower level the most prominent vein is 3 to 4 inches wide, with an ore course varying from one-eighth to one-fourth inch in width and is frozen to the walls. It strikes N. 10° E. and dips 66° S. 80° E.; the other veins differ somewhat, both in dip and in strike.

The ore body is a typical filled fissure. It varies in width from about 2 inches to 4 feet at the widest points, with an average of 1 foot. The dip is fairly uniform, being 70° SE. The vein is formed along two series of intersecting fractures, which strike N. 10° E. and N. 35° W. No one of these fractures is very long and the main open space extends along one fracture until this becomes narrow, and then passes to a fracture of the other strike, to be again taken up by a fracture parallel to its first course, and so on. This feature is well shown in figure 5 (p. 42). It is probable that similar variations could be observed in dip, but exploration is not yet sufficient to disclose them. The fissuring has been accompanied by very extensive brecciation, so that the vein is characterized by many angular inclusions of country rock, by innumerable branches, and by other irregularities. (See fig. 15.) In places it is merely a sheeted zone which has but little open space for ore deposition.

The structure of the vein filling appears massive on first sight, probably on account of the great mass of highly altered rock with which the fissure is filled, but on closer examination very fine druses are often discernible, either parallel to the walls or following around the periphery of included fragments.

Replacement of the country rock has been more extensive in this vein than in most of those in the region. Here, as elsewhere, however, it has been chiefly confined to silicification, which has transformed the included rock fragments into hard black flinty silica. Barite has also been to some extent introduced into the wall rock, and included fragments show a higher degree of replacing action than common.

The accompanying sketch shows the brecciated nature of the vein.

Ores.—The vein filling consists of the following minerals, named in the order of deposition: Primary minerals, black dense silica, barite, galena, freibergite, barite, white quartz, pyrite, and sphalerite; secondary minerals, pyrargyrite and secondary galena. Silica and barite are the most abundant.

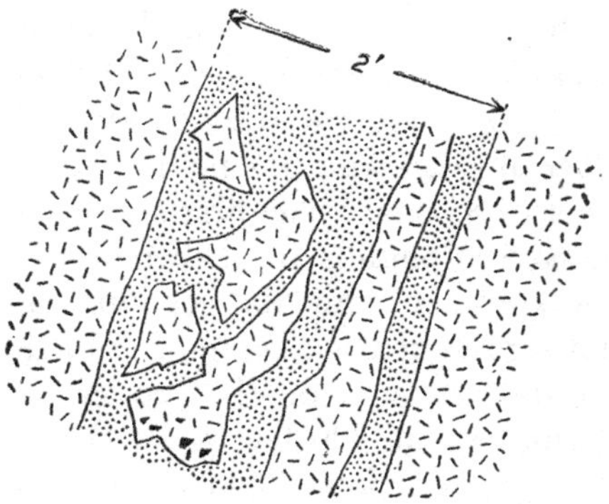

FIGURE 15.—Face of Pelican drift, showing structure of lode.

A little rhodochrosite also occurs in subordinate amount. A number of these minerals, especially barite and quartz, probably appear in more than one generation, but the relative periods of deposition can not in all cases be made out with certainty.

The secondary pyrargyrite and secondary galena are without doubt connected genetically with oxidation processes. The pyrargyrite occurs in later crevices in the ore as well as in defined crystals lying on the outer surface of the white quartz druses in open cavities. It has, with little question, originated from the reprecipitation of antimonial silver solutions derived from the alteration of the freibergite (p. 63). It is probable, for this reason, that the ruby silver will not continue in depth, or at least will occur only as far as surface solutions could gain access along especially favorable channels.

The ores contain many irregular vugs, some of them 8 inches or more in diameter. Nearly all of them are lined with white quartz, usually in fine druses. Barite crystals project into many of them, but all of these are covered with a fine coating of white quartz, showing that silica was the final phase of the deposition. What is probably secondary galena occurs coating the surface of the sphalerite along cracks and fractures in the ore. In the hand specimen it may be seen merely as a dull lusterless surface, slightly grayish in appearance, in sharp contrast to the brilliant luster of the fresh fracture of the sphalerite when this mineral is broken, where no crack large enough to permit the entry of surface solutions has been available; under the microscope, however, it is distinctly coated with some gray mineral. In some massive sphalerite the minute galena-coated crevices are so numerous that it is nearly impossible to break the ore so as to get a perfectly fresh fracture.

MISSOURI FAVORITE.

The Missouri Favorite mine is on an unpatented claim located just north of Henson Creek, about three-quarters of a mile west of Lake City. The mine is one of the most accessible in the region, being near the railroad terminus and a source of supplies.

Production and development.—Little is known regarding the history of the mine. Mr. Snyder of Lake City has worked the property for a number of years and has made two shipments of ore, one of 10 and another of 8 tons. The first shipment gave returns of $8 per ton and included 800 pounds of sacked ore, which ran $247 per ton. The second shipment ran $50 per ton. These are picked lots of ore and do not represent the average run of the ore body.

The development on the property consists mainly of drifts. One drift runs about 600 feet from the mouth of the adit tunnel along the main vein. Some 250 feet in a stringer, making an angle with the main drift of 20° to 30° to the northwest, has been followed for 350 feet.

Veins.—The vein is in Eureka rhyolite, which shows more alteration than is common in the area. In general the localities showing the greatest amount of alteration and decomposition appear to have been the least prospected, and, indeed, have been left practically unexplored on any extensive scale. The cause for this is not apparent, although it may be due to the fact that nearly all of the mining operations in the Lake City district have been either in connection with placers or on well-defined quartz veins, and the prospector is loath to take up a new kind of exploration.

The faulting in the vicinity of this vein is more noticeable than it is near many of the mines visited. Most of it seems to have occurred on the west side of the vein, where great slickensided surfaces are common. Gouge is present in enormous amounts and during wet seasons of the year clay and mud fall from the roof in tons. The hanging wall is in several places separated from the footwall by a great wide fissure filled with clay and mud. As the portion of the vein developed is near the grass roots, much of the clay and mud may be due to surficial wash. Toward the west of the main working an enormous horse comes in and separates the two walls by 30 or 40 feet.

From the foregoing it may be inferred that the fissure is irregular and shows great variations in size and character. The strike of the vein is about N. 50° W., although there are so many variations that only an approximate direction can be given. The dip is in general about 70° SW. The width varies from a few inches to several feet.

Ores.—The ore minerals are galena and tetrahedrite with some chalcopyrite and sphalerite. The gangue is composed almost entirely of barite with a little quartz. The ores show banding, as is common in the ores from this area. The alternate layers of decomposed rock, barite, and ore minerals resemble parts of the Ute-Hidden Treasure vein.

SAN CRISTOBAL GROUP.

MONTE QUEEN.

The workings of the Monte Queen mine lie due south of Lake City, on the west side of Lake Fork, on the eastern slope of the mountain, 9,100 feet above sea level and 300 feet above the bed of the river. The mine is connected with the main road by a short branch road which makes it easily accessible.

Vein.—The vein is opened on its outcrop by a tunnel driven S. 73° W. for 950 feet. The country rock is a grayish stratified breccia evidently formed by the fracture of the cooling upper surfaces of successive lava flows during their extrusion. Some stoping has been done in the mine, but no records of its production or history could be obtained.

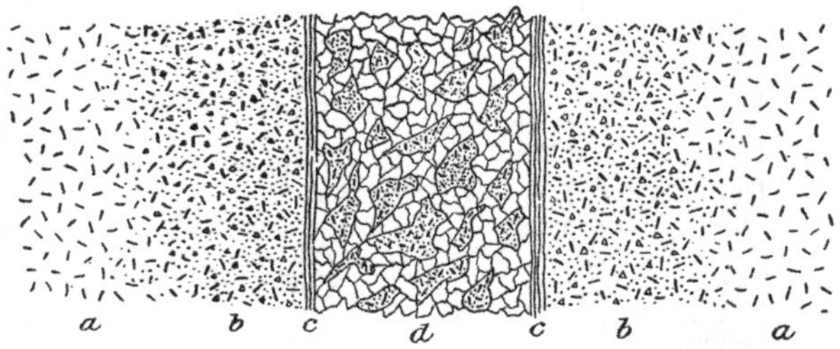

FIGURE 16.—Cross section of the Monte Queen vein, showing included rock fragments, strong selvage, and massive vein filling. *a*, Unaltered country rock; *b*, silicified flow breccia; *c*, selvage; *d*, fragments of country rock and barren quartz gangue.

The workings are on a fissure vein, produced in greater part by the filling of an open crevice and only in minor degree by replacement of the country rock. The vein, which departs but little from the vertical in its explored portions, is about 3 feet in width. In some places it has well-defined walls, locally separated from the country rock by strong selvage seams. (See fig. 16.)

Ores.—The ore minerals are segregated within the barren gangue material into vertical shoots which show no appreciable pitch within the plane of the vein (see fig. 17). Between the shoots the vein

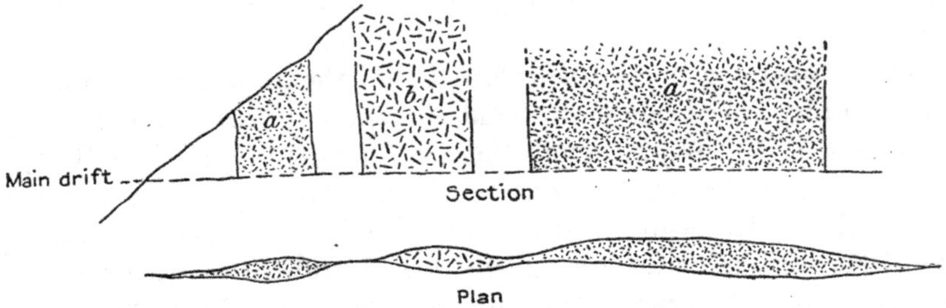

FIGURE 17.—Plan and longitudinal section of Monte Queen vein, showing ore shoots. *a*, Tetrahedrite and pyrite; *b*, galena and zinc.

pinches so that only a few sheeting planes with stringers of quartz, etc., are left, by which the vein may be followed from one shoot to the next. The shoots have a stope length of 50, 63, and 200 feet, respectively. Nos. 1 and 3 contain chiefly pyrite and some gray copper ore and do not run very well in silver. No. 2 consists chiefly of a peculiar dark-brown, fine-grained, massive, rosin-like sphalerite, mixed with fine steel galena and averaging about 75 ounces in silver and notable percentages of bismuth. Some of this ore, according to the management, ran as high as 2,000 ounces in silver and contained 20 per cent of bismuth. The pyritic shoots were too low grade to pay. A fourth shoot, or widening of the vein, 5 feet in width, occurs in the breast of the tunnel 950 feet west of the mouth; so far as yet explored it contains nothing but massive rhodochrosite.

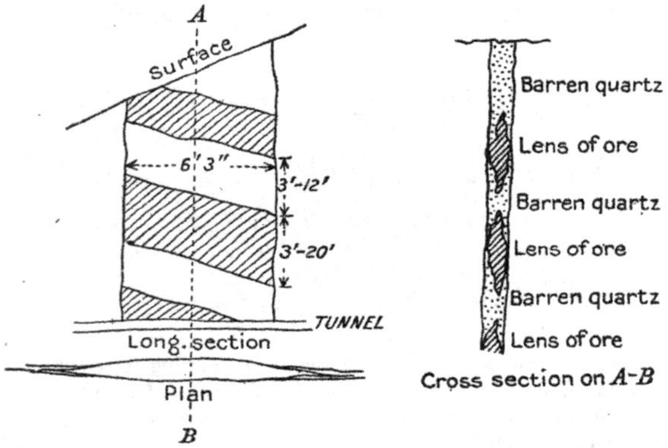

FIGURE 18.—Occurrence of shoots within shoots, Monte Queen vein.

The metallic minerals in the No. 2 shoot occur in smaller shoots, which are separated by vertical intervals of barren quartz and are lenticular in both vertical and horizontal section; they pitch slightly (15°) along the course of the vein to the southeast. The vertical extent of these lenses is from 3 to 20 feet (see fig. 18) and that of the barren quartz filling between from 3 to 12 feet. The barren portions are filled with included and highly silicified fragments of wall rock and the country rock is altered to distances of 4 to 5 feet from the vein filling.

NELLIE M.

The workings of the Nellie M. mine are west of those of the Monte Queen. The vein is opened by an upper and a lower drift. The upper tunnel, which is 510 feet in length, runs S. 53° W. with the vein. The lower tunnel is 700 feet in length. The country rock is the usual bedded type of andesitic and rhyolitic flow breccia, much decomposed, and with very prominent banding. The dip is 65° S. The vein, so far as could be learned, has not yielded ore in commercial quantities, but a little sphalerite, chalcopyrite, and some silver have been found. A few streaks of metallic mineral could be seen here and there along the broken sheeted zone followed by the tunnel.

LODE STAR

The Lode Star mine workings consist of a shallow shaft and a few short tunnels located on a shear zone from 2 to 300 feet wide. This zone has had so little work done on it that its nature can not be determined. The mineralized zone is greatly stained with iron oxide, is much silicified, and is reported to have yielded some values on assay, but has so far not paid for what exploration has been done.

DAUPHIN, ROB ROY, AND SULPHURET.

On the east side of Lake Fork, about 1¼ miles south and a little east of Lake City, is a great shear zone striking about N. 73° E. On one side of this are the workings of the Rob Roy and Dauphin and on the other to the southeast are those of the Sulphuret. Sufficient work has been done to disclose some gray copper ore, several carloads of which have been shipped from the Sulphuret; but the development is insufficient to furnish any reliable data. It is possible that the Sulphuret may represent the northeast continuation of the Monte Queen vein.

GOLDEN WONDER.

The geologic map (Pl. III) shows an area of silicified rhyolite covering a large part of the southeastern portion of the Lake City quadrangle. Within this area, some 2¼ miles southeast of Lake City, is the Golden Wonder mine. This property is unique in that it is the only deposit of those examined by the writers in this area that can be classed as a true replacement deposit.

The Golden Wonder claim is on the northern side of Deadman Gulch, its side lines extending east and west, approximately parallel to the trend of the gulch. A good trail connects the workings with the main wagon road to Lake City. Underground examination of the Golden Wonder was not possible because of the condition of the workings, but a basis for conclusions as to its deposits was supplied by the examination of contiguous and doubtless analogous deposits.

Few authentic data are available in regard to production. Two carloads of ore are reported to have been shipped in 1906 which are said to have averaged $70 a ton in gold.

The original discovery was made at an elevation of 10,500 feet, at which height an adit tunnel is said to have been driven 150 feet east into the silicified country rock. Another adit tunnel, started at an elevation of 10,375 feet, near the center of the gulch, is said to have been driven 850 feet. This crosscut is not a direct one, but is reported to twist and turn in a remarkable manner.

The country rock in this vicinity is an altered rhyolite, doubtless of intrusive origin. This rock has been greatly changed by silicification and in places by sericitization. Solutions containing pyrite have impregnated the surrounding rocks for several miles in all directions, and it is probable that in these is to be found the source of the gold found in the Golden Wonder workings and elsewhere in the vicinity, local concentrations of the gold contained in the pyrite having produced the few rich pockets found in the vicinity.

True quartz veins are scarce in the workings examined on contiguous properties, the mineralization appearing to be a simple replacement of the country rock by silica and pyrite. In places, however, mineralization, presumably by secondary concentration, has taken place along joint planes and fracture zones in the altered rhyolite.

MAYFLOWER AND CONTENTION CLAIMS.

The Contention mine includes the Contention and Mayflower claims, which are located about 1,000 feet north of the north end of Lake San Cristobal and 3 miles south of Lake City. The claims are so located that the west end of the Mayflower adjoins the east end of the Contention.

Production and development.—For short periods during its history the mine has produced a considerable amount of silver from its upper workings. The ore bodies were small, however, and the mine has not at any time been capable of maintaining a heavy output. For the four years, 1889 to 1892, the following statistics on the Mayflower and Contention are given:

Production of Mayflower and Contention claims, 1889–1892.

	Gold.	Silver.
1889	$120	$905
1890	63	1,244
1891	....	2,456
1892	47	3,543
Total	230	8,148
Grand total	$8,378	

VIEW LOOKING SOUTH UP LAKE FORK.

Lake San Cristobal shows in the distance. Contontion Mill is on tho right. Photograph by Whitman Cross.

The workings are located on the west side of Lake Fork just above the wagon road from Lake City. The tunnel openings by which the vein is reached lie on the extreme north end of the Mayflower claim, but most of the work is on the Contention claim. (See sketch, fig. 19.)

The mine is opened by three tunnels, two on the Mayflower claim and one on the Contention. The lower tunnels follow the vein westward into the hill and show low-grade ore. The upper tunnel and the workings connected with it are now abandoned but show considerable stoping and indicate that quite a little ore has been removed.

A large mill for treating the ore was erected in 1904. From the Gates crushers the ore was sent to rolls, thence to screens and to a Huntington mill, thence to classifiers, and after that to Wilfley tables. A high-grade concentrate was obtained, and the management claims to have saved 90 per cent of the gold and 88 per cent of the silver.

Country rock.—The country rock is the Picayune volcanic group. The ore body is a fissure vein with a nearly vertical dip, with a slight northward tendency, and a somewhat variable trend, averaging approximately S. 60° W. to S. 79° W.

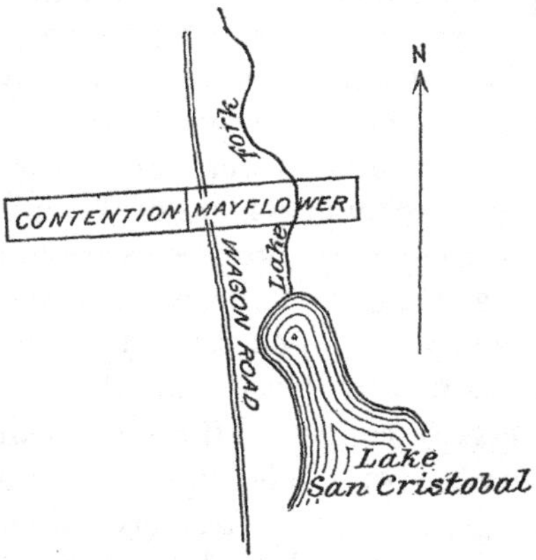

FIGURE 19.—Sketch of Mayflower and Contention claims.

Vein.—The vein, which has an average width of from 14 inches to 2 feet, is in places single but not uncommonly splits into stringers along the strike. In the lower tunnels it is more irregular than in the upper tunnels and stopes, from which practically all of the merchantable ore had been taken at the time of examination in 1904. Pinches and swells are numerous, the vein having practically no width between stopes. Banding is generally imperfect or absent except in small quartz offshoots, where comb structure is better developed.

Ores.—The ore consists of silver-bearing gray copper or freibergite, (slightly auriferous), sphalerite, chalcopyrite, and pyrite in a gangue of quartz, barite, and rhodochrosite.

In the upper workings the gray copper predominates and often makes up the bulk of the ore with a little rhodochrosite, but in the lower tunnels pyrite is in greatest abundance and gray copper of only secondary importance. Some of the upper ore was extremely rich, $60,000 worth of silver having been extracted from a single small pocket. Barite and quartz are less abundant in this upper ore.

The ore from the lower tunnels is a massive aggregate of dense, fine-grained chalcedonic silica containing innumerable large, open vugs. In this quartz great numbers of thin tabular crystals of barite form a network which in many places resembles the feldspars in a coarse diabase. Mixed through this quartz-barite gangue is fine-grained noncrystalline pyrite, sphalerite, and tetrahedrite, with very subordinate galena and chalcopyrite. Open vugs are lined with beautiful crystals of barite generally coated with druses of quartz. Except that the quartz which coats the barite crystals is the last formed, the relative ages of these different minerals can not be definitely determined. In much of the ore secondary pulverulent chalcocite has been deposited as a secondary sulphide on the surface and in the cracks of the pyrite crystals. The wall rock has been very much altered by the vein solutions and shows a white bleached appearance for more than a foot beyond the boundary of the vein matter. This bleached rock has been extensively sericitized and is heavily impregnated with cubical pyrite. The gray copper and other metallic minerals do not extend beyond the zone of the main vein mass.

The ore occurs in shoots, between which the vein pinches to a narrow sheeted zone, which is slightly replaced by silica and a little pyrite, but has afforded no open space for deposition. The upper and lower workings were not connected at the time of the investigation, and the manager was of the opinion that they are not on the same vein, owing to the greater preponderance of the gray copper in the upper workings; in the writers' opinion, however, this difference is due simply to the customary gradual falling off in high-grade silver minerals with increasing depth.

The gray copper is the most valuable mineral normally found in the ore body; in the upper or more oxidized portions of the old workings, however, extremely valuable pockets of ore carrying native silver are reported to have been found. The gray copper carries high values in silver, and where massive often assays as high as 1,000 ounces to the ton. The pyrite in the lower ore carries gold, and a test sample of 2,500 pounds is reported to have yielded 0.28 ounce of gold and 11.34 ounces of silver.

GOLDEN FLEECE.

The Golden Fleece mine contains one of the celebrated veins of Colorado, and it is much to be regretted that the condition of the workings has made an exact and satisfactory examination of the mine impossible. A large portion of the upper workings is now inaccessible, no faces of ore remain in what few stopes can be observed, and the nature of the ore from the most productive portion of the mine can be determined only from specimens preserved in the offices of the company or casually picked up on the old dumps of the mine. The description of the mine here given, therefore, leaves much to be desired, though it is hoped that it may prove serviceable to those who are interested in the property.

Location and history.—The upper and productive workings of the mine (see Pl. VII) are located about 2,600 feet west of the north end of Lake San Cristobal, on the easterly slope of a flat-topped, mountain 11,800 feet high. The workings are about 10,000 feet above sea level and 1,000 feet above the level of the lake. (See fig. 20.)

The outcrop of the vein is located in a broad gulch that runs with the slope of the hill. It trends about N. 60° E., forming a promi-

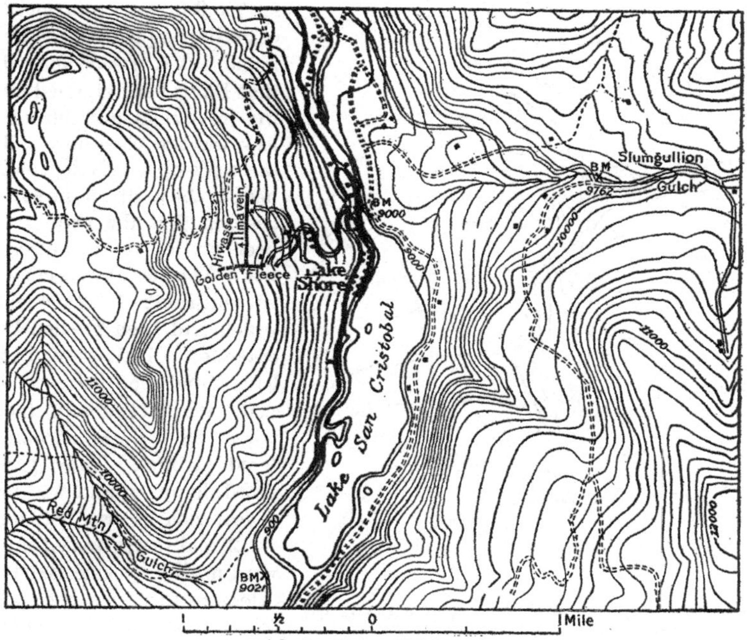

Contour interval 100 feet.

FIGURE 20.—Topographic map of the region around Lake San Cristobal.

nent ridge that rises sharply from the broad bottom of the draw to a height of 100 to 150 feet with a width of 20 to 30 feet. (See fig. 21.)

The mine is reached by a good wagon road from the lake, and transportation of ore to Lake City has offered no serious difficulties.

The history of the Golden Fleece and Black Crook mines is interesting, and the following extract has been taken direct from an article by Rickard:[1]

In 1874 Enos F. Hotchkiss, connected with a Government surveying party which was laying out a toll road from Saguache to Lake City, caught sight of the outcrop of the Golden Fleece standing conspicuously above the hill-slope, and examined it. He located it as the "Hotchkiss" mine and had some assessment work done while

[1] Rickard, T. A., Across the San Juan Mountains, Eng. & Min. Jour., vol. 76, 1903, pp. 307–308.

he was engaged in his survey work in the vicinity. As far as known he found no ore. A year later, when Hotchkiss had abandoned his claim, it was re-located by George Wilson and Chris Johnson, under the name of "The Golden Fleece." They began what is now known as the No. 1 tunnel, but finding only little stringers of rich ore, they ceased work. Others did similar desultory prospecting. O. P. Posey found a very rich bunch of ore in the croppings above the No. 1 tunnel and took out several hundred pounds, which were packed to Del Norte and sent thence to the Pueblo smelter. Then John J. Crooke took a lease and bond; he also extracted about $30,000 from the outcrop above No. 1 tunnel, which had been extended a little farther, without result. This was between 1876 and 1878. In 1889 Charles Davis took a lease and bond; he did a good deal of work along the high croppings, and finally sunk a shaft 30 feet deep, which struck a body of ore yielding $40,000 in a very short time. Later in that year, 1889, George W. Pierce bought the mine for $50,000 and commenced extensive explorations. He found out very soon indeed that Davis had extracted all the ore in sight, and the outlook was not cheerful. All the work up to this time had been to the north on the supposition that the vein had been faulted in that direction. The new owners crosscut south at the No. 2 tunnel, which had been previously extended a little way, but had found nothing. The vein was picked up, but not much ore was encountered at first. They persisted, however, and within a year rich ore was cut on No. 2, and it was traced upward until it became easy to intercept the same body at No. 1. It was discovered that the former owners had been within 10 feet of the main ore body of the mine, which from that time until 1897 was very profitable.

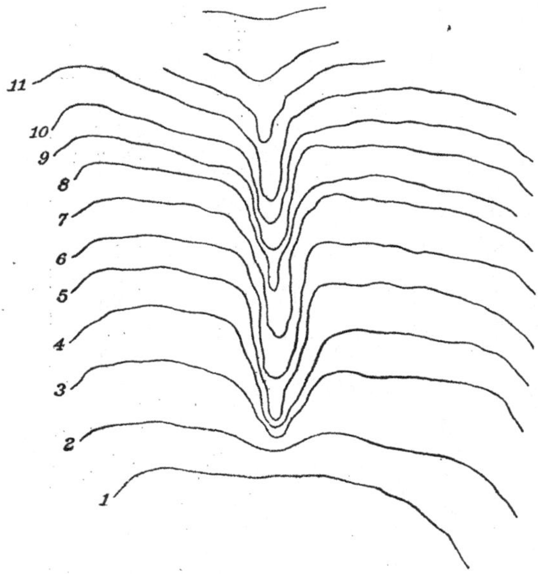

FIGURE 21.—Sketch of ridge formed by the outcrop of the Golden Fleece vein. (Not to scale.)

Nearly all of the ore of merchantable grade produced was taken from the stopes above the third level. A few bunches of high-grade ore were found down as far as the main tunnel, where one bunch gave the phenomenal assay of 125 ounces gold and 1,255 ounces of silver. This was very exceptional, and no ore has been found in quantity below tunnel No. 3.

The rich ores of this mine did much to stimulate active prospecting in this region, and it is probable that to it are indirectly due many of the other discoveries in the district.

Development and production.—The vein is opened by four tunnels. Two of these are drifts on the vein and are located on the crest of the ridge made by the outcrop. Below is a short crosscut tapping the vein at a slight angle. In and above these three upper levels are located the large stopes from which nearly all of the rich ore of the mine was taken at a maximum depth of not more than 400 feet from the surface. A shaft connects with the upper level west of the mouth. From the lowest of these three levels the vein was for some time worked through winzes, but later a long crosscut tunnel was driven 1,200 feet below the collar of the shaft. This intersected the vein and intermediate levels have been worked by means of it. (See fig. 22 and Pl. VII.)

According to Mr. George W. Pierce the mine produced $1,400,000 up to 1904.

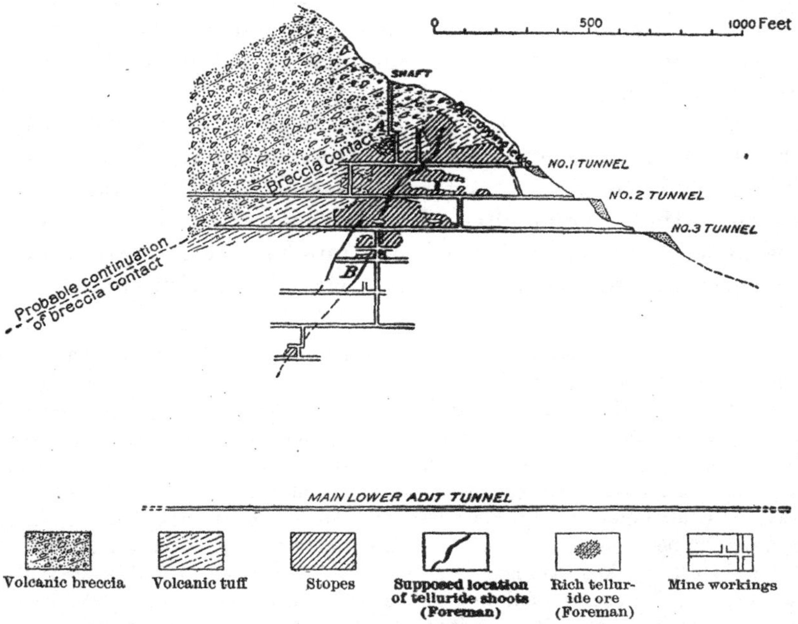

FIGURE 22.—Longitudinal section of the Golden Fleece vein, showing position of agglomerate contact and ore shoots.

Country rock and structure.—The country rock of the Golden Fleece consists of a series of plainly stratified flow breccias interbedded with volcanic tuffs and agglomerates. The planes of stratification are peculiarly well seen in the weathered rock, which is whitish and decomposed into a loose friable clayey material. In the Ilma workings, north of the Golden Fleece mine, these tuffs and breccias dip east, but in the Golden Fleece they are reversed to a strong westerly dip of 27°, which prevails through both surface workings and in depth.

Above the highest tunnel the fine-bedded tuffs and flow breccias give place to a conformable series of extremely coarse agglomerates containing subangular bowlders embedded in a yellowish tuffaceous matrix. Many of these bowlders are several feet in diameter and constitute most of the rock, the matrix being greatly subordinate. They are composed of the same material as the finer tuffs and breccias below and in many places show well-marked banding, set at all angles according to the position of the bowlder at the time of deposition. This coarse agglomerate, which is very thick, was traced up the hill to beyond the upper tunnel, but its total thickness was not determined.

The contact of this agglomerate with the underlying beds is conformable, but, as explained below, is probably also a bedding fault. The vein throughout all of the upper workings, where explored, terminates against this contact as a roof and nowhere enters the agglomerate itself. A tunnel was driven across the contact into, though not through, this agglomerate, disclosing neither a different type of rock nor any evidence of mineralization.

The abrupt termination of the Golden Fleece vein against this plane has been the cause of much expensive exploration and has been explained in different ways by different observers. Rickard believes that owing to the different physical character of the coarse breccia and the underlying tuff the vein broke into minute stringers and never penetrated the coarse breccia. His description warrants quotation:[1]

> The outcrop ceases when the vein encounters the coarse breccia; so, also, in the underground workings the vein itself comes to an end with a suddenness which is, however, only comparative. The contact has been considered a fault; a good deal has been said concerning its regularity and clean-cut character. This, however, does not, I believe, accord with the facts. The so-called fault is not a break or dislocation in the rocks; it merely marks the division between the layers of fine-grained breccia and an overlying formation of very coarse breccia. There is no smooth plane or wall or defined parting between these two formations, but only a sudden transition, which at a distance is more marked than near by. * * * The contact existed before the vein was formed. The fracture, followed by the ore, passed easily through the finer-grained rock, but ceased abruptly when it met the beds of coarse breccia, because the force of fracturing was not only insufficient to overcome the resistance of the harder fragments contained in the latter, but it must have been dissipated by the encounter with a loose-textured body of rock, much in the way that the power of a diamond drill becomes wasted in passing into a shifting mass of loose conglomerate. As a consequence the energy of shattering was diverted along the contact, the vein fracture ceased, and the later ore-depositing waters were barred from further advance into the coarse breccia, save as a scattering confined to the neighborhood of the contact. At the third level the ore body, occurring in the fine-grained country, was notably wider immediately at the "contact," and in examining the outcrop of the vein I noticed that it was difficult to decide upon the exact line of separation between the two formations, because the mineralization extended from the fine into the coarse breccia so as to obscure the divisional plane.

---

[1] Op. cit., p. 346.

The writers can not agree with Rickard as to the nature of this plane. They were unable to note any points at which the ore actually entered the coarse breccia, but did, on the contrary, detect fragments of ore material included in the breccia along the line of division between the two series. In their opinion, therefore, the line of demarkation between the coarse breccia and tuff has been also the locus of a bedding fault which has occurred since the vein formation and has disturbed the continuity of the vein. Whether exploration will reveal the westward extension of the Golden Fleece vein in the coarse agglomerate or not can not be said, and it is doubtful if the different character of rock would permit its continuation with any degree of regularity, but the writers believe that the faulted portion exists and may yet be found.

Vein.—As far as could be determined from the exposures seen and the specimens obtained, the vein of the Golden Fleece mine consisted of a mass of very irregular broken country rock with interstices filled with dense granular gray and white quartz and fragments intensely altered to a hard fine-grained silicious rock impregnated with petzite, tetrahedrite, and other minerals. This ore formed an irregular zone 8 or 10 feet in width, which from the superior hardness imparted to it by mineralization stood out in prominent outcrop above the surrounding rock. This outcrop was stained brilliant yellow, with here and there a reddish cast, both probably due to the presence of oxidized iron minerals and to the kaolinization of the clayey material in the tuffs. Some brick-red stains are also believed to be due to the presence of tellurous oxide.

The outcrop is now honeycombed by the work of leasers gouging around for the many small patches of high-grade ore found near the surface. In the upper levels the vein seems to have had no definite walls, but to have simply faded out into less broken and unmineralized rock. In the lowermost levels (main tunnel level) it is a narrow but distinct filled fissure with fairly definite walls. The vein here could be clearly seen only in the face of the long tunnel, and consisted chiefly of light pink rhodochrosite, containing comparatively little metallic mineral.

From a wide broken zone at the surface the vein, as stated, narrows down in depth to a fraction of a foot. Between the main or adit tunnel level and the third level it could not be seen, and therefore its character could not be determined.

The vein in the lower levels shows the character of most of the veins of this region; that is, instead of remaining a single fissure it splits into a number of smaller fissures which separate more widely in depth and branch out from each other when followed along the strike. (See fig. 23.)

Ores.—The ore from the Golden Fleece mine consisted of petzite, pyrite, argentiferous tetrahedrite, galena, hinsdalite,[1] and pyrargite in a gangue of fine-grained gray and white quartz and some rhodochrosite. As the mineral hinsdalite is new, its association with telluride ores is interesting. The gray color of the gangue is in places due to the presence of minutely divided grains of petzite.

Of the metallic minerals, petzite is the most important and carried the chief gold values of the mine. Galena was scarce and gray copper fairly abundant. The latter mineral offered something of a contrast to its occurrence in other Lake City veins, as it rarely, if ever, contained more than $60 worth of silver per ton.

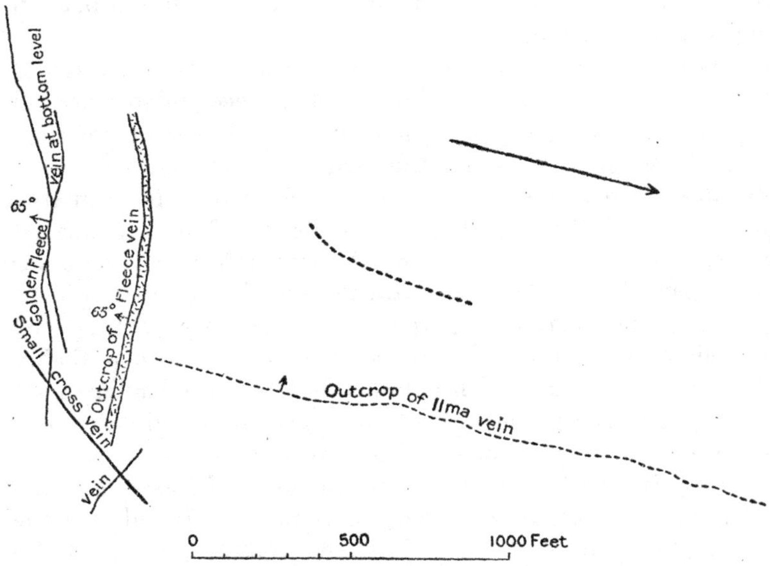

FIGURE 23.—Outcrop and branches of Golden Fleece vein.

The ruby silver came in irregular bunches, some of which are reported to have been found 1,200 feet below the surface; but they were very much more common in the zone of sulphide enrichment and, in the writers' opinion, are clearly of secondary origin. (See pp. 62–63.)

The proportion of gold to silver in the ore may be seen from the following record of carload shipments furnished by the management:

Precious metals in ore of the Golden Fleece mine.

Class of ore.	Gold.	Silver.	Ratio by weight.	Ratio by value.
	Ounces.	*Ounces.*		
1	134.10	3,077.0	1:23	2:1
2	6.11	238.0	1:40	1:1
3	2.0	53.0	1:26	10:7
4	0.40	15.0	1:37	1:1

[1] Hinsdalite is a new mineral described more in detail on p. 54.

Apparently the richer ore had a slightly higher proportion of gold, but in general the ratio of silver and gold by value may be said to have been roughly 1:1.

Most of the rich ore of the Golden Fleece mine was shipped to the smelters, but the low-grade mill stuff was treated on the spot. As the values were chiefly contained in telluride minerals (principally petzite, but also some hessite), the treatment, by concentration, presents features of interest. The mill was of latest design, erected by Stearns, Roger & Co. It consisted of rolls for crushing, Huntington mills for regrinding, Wilfley tables for concentration, and a canvas plant for slimes. No use was made of amalgamation. The Huntingtons were provided with screens of 30 mesh, and experience showed later that 20 mesh would have been better. In treating 18,000 tons having an average assay value of $10.25, half of which was in gold and half in silver, the extraction averaged between 45 and 60 per cent; 63 per cent was the best result. The concentrates contained 55 to 65 ounces of silver, 1 to 3 ounces of gold, and 12 to 18 per cent of lead, in the form of galena. The concentration was in the ratio of 12 to 1.[1]

The ore in the vein (see fig. 23) occurred in a shoot measuring about 750 feet along the levels. As it was followed down, this shoot narrowed to a point below which only isolated bunches of ore were found. Within this shoot a more or less central interior shoot of unusually rich telluride ore is reported to have occurred, as indicated by the wavy black lines in figure 22. An extremely rich bunch of telluride was also found above the first level, west of the shaft, adjacent to the contact of the overlying volcanic agglomerate.

It is notable that this ore shoot occupies nearly the position that would be taken by the trace of the intersection of the Ilma and Golden Fleece veins, but as this intersection could not be actually observed it is not possible to define its effect on the production of the shoot.

The depth of oxidation as affecting the transformation of tellurides into the native metals does not seem to have been great, but as no observations could be made in the upper stopes this must remain uncertain. Secondary enrichment, though apparently without effect on the tellurides, has undoubtedly led to the formation of the rich masses and bonanzas of ruby silver found here and there throughout the mine, presumably by the solution of the tetrahedrite and its later precipitation as the richer mineral pyrargyrite. The fact that bunches of this ore were found 1,200 feet below the surface shows that the surface water has been able to penetrate along cracks and crevices to considerable depth. The ore has without doubt originated chiefly from a replacement of the country rock and a subordinate amount of actual cavity filling.

BLACK CROOK.

The Black Crook mine is on the eastern slope of the mountain which lies 2,400 feet west of the north end of Lake San Cristobal, north of and adjacent to the upper workings of the Golden Fleece.

[1] Rickard, T. A., op. cit., p. 346.

It is approached from the main wagon road in the valley of Lake Fork by a wagon road which winds upward over the hill. The mine was operated intermittently for perhaps 12 years until about 1903.

Development.—The outcrop of the vein runs along the brow of the hill 1,264 feet above Lake San Cristobal, bending westward more than the strike, owing to the westerly dip and the curve of the hill. The mine is opened by crosscut tunnels. Two of these run slightly north of west, tapping the vein near its southern end. The highest, known as the Capel crosscut, cuts the vein, which runs S. 30° W., at 150 feet below the outcrop at a slight incline. An incline shaft connects this crosscut with the lower levels, and through this the mine has been worked to a depth of 820 feet. From the upper workings profitable stoping has been carried on.

As in the Golden Fleece, Vermont, and many other mines, the difficulty of handling water and the added expense of mining at depth led to the driving of a long adit tunnel. Part of the long tunnel of the Golden Fleece mine adjoining this property on the south served for both mines. From this adit at a point 1,950 feet from the mouth, a branch known as the De Camp crosscut was driven, tapping the Black Crook vein 1,200 feet below the outcrop. This tunnel was then connected with the upper workings by a raise on the vein. Below this tunnel level a winze 100 feet deep was sunk and 240 feet of drifting was done. The results obtained by this deep and expensive work have been uniformly disappointing in this, as in the other mines of the district. The amount of drifting and stoping may be seen from the map of the workings (Pl. VIII) and the longitudinal section (fig. 24).

Country rock.—The country rock, in which the Ilma and Gold King veins of the Black Crook mine occur, belongs to the same series of bedded flow breccias and tuffs that are found in the Golden Fleece mine. The beds dip 15° to 40° E. and the veins intersect them at right angles. (See fig. 25.) In the upper levels the tuff is weathered into a yellowish-white banded rock whose stratified nature is so well developed as to strongly resemble sandstone. The beds vary from half an inch or less to 5 or 6 inches in width. None of the coarse agglomerate noticed in the Golden Fleece mine was seen in the Black Crook. Interstratified with the layers of tuff are denser though often equally well banded flow breccias, which when noted in the deeper mine workings are greenish or brownish gray in color and can not easily be confused with the tuff. These flow breccias are much more extensively developed on the Golden Fleece tunnel level than in the upper workings of the mine.

Veins.—The Black Crook mine owes the major part of its production to a single vein known as the Ilma vein. Other minor veins and branches have been somewhat explored, and a little ore has been found along the Black Crook fault, but its amount has been trifling. The Ilma vein strikes nearly north and south; it shows many local deviations, and these are often sharply angular, as is the case with nearly all the Lake City fissures, but the average direction is remarkably straight.

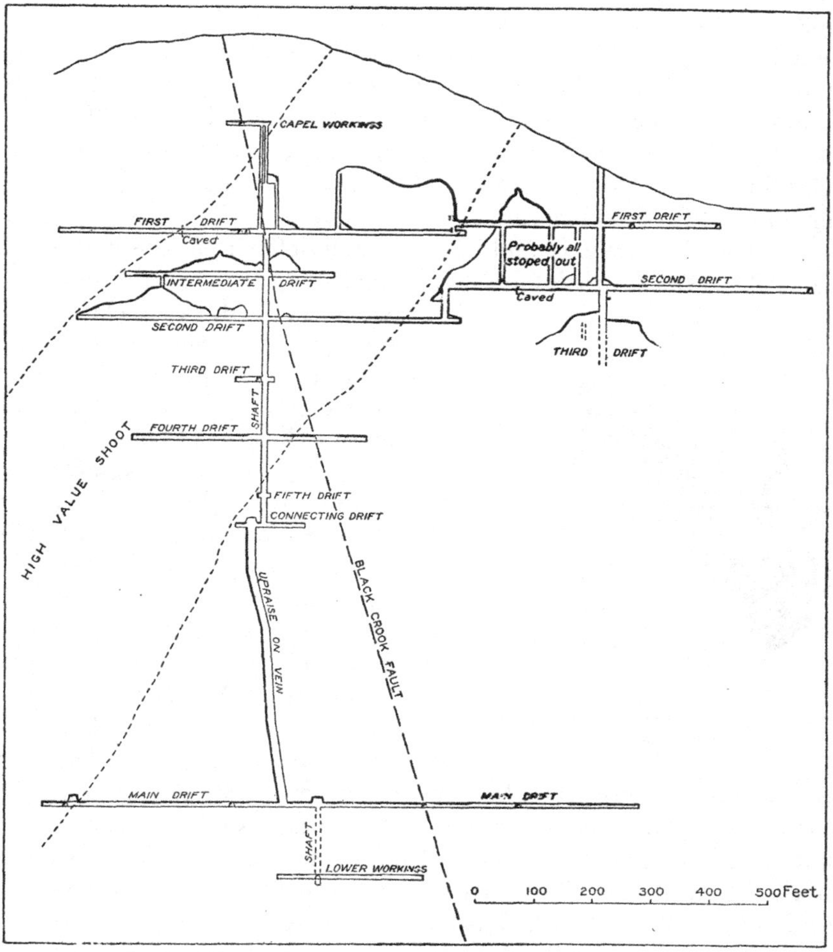

FIGURE 24.—Longitudinal section on plane of Black Crook vein.

The dip is toward the west. At the surface it was 70°, shallowed to 50° and 45°, then steepened again to 58°, which is its inclination in the lowest workings, 1,300 feet below the surface.

The vein has numerous branches, most of which make out from the hanging wall. (See fig. 25.) The foot wall is much more regular and is frequently separated from the vein filling by a strong layer of selvage clay; this, however, is not constant. The vein is not a simple filled fissure, but rather a series of sheeting planes, in many places brecciated, which have been replaced by ore and gangue minerals so as to form what Ransome has termed a lode fissure. From one extreme, where all ore shows evidence of formation by replacement, to the other, where a narrow and well-defined filled fissure occurs with no discernible replacement, this vein furnishes examples of nearly all gradations.

The mineralized zone has a width of from 4 to 8 feet through much of the mine, but only in the older and more profitable stopes did this width pay for extraction, and then only in exceptional cases. The pay streak in the ore, where the mineralization had been most intense and the filling least confused by waste rock, averaged throughout the mine about 18 to 20 inches.

Fault.—The Ilma vein is faulted by a well-defined east-west dip fault, called the Black Crook fault on account of its having been first observed on the Black Crook claim. This fault dips 73° W. and has a reverse throw as it offsets the north end (the hanging-wall side of the fault) of the Ilma vein about 35 feet to the west on each successive level. The fault is evidently later than the vein, as it is mineralized only where secondary oxidation and enrichment processes have been active since surface alteration of the vein began.

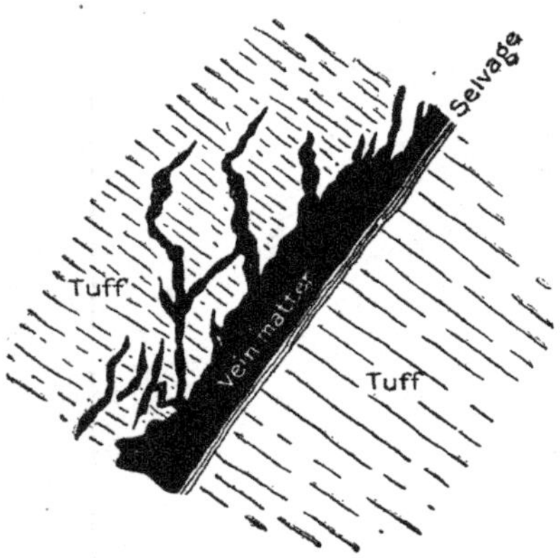

FIGURE 25.—Type of fissure in Black Crook (Ilma) mine.

Ores.—The ore of the Black Crook consists of sphalerite, galena, a little tetrahedrite, pyrite, and very subordinate chalcopyrite with irregular bunches of pyrargyrite in a gangue of quartz, barite, and rhodochrosite. The pyrite usually is in small amount in the ore, though it frequently impregnates the country rock and is then present in well-defined cubical crystals. The most constant mineral is sphalerite, which is but rarely absent in any of the ore and in some stopes is so abundant that the term "zinc stope" has been applied to them. It is usually of the coarse cleavable variety and contains enough iron and impurities to make it rather dark in color. An average of 158 samples taken on all levels of the mine showed 3.20 per cent lead and 15 per cent zinc; this gives a fair idea of the general content of the ore in sphalerite and galena. Dark ruby silver, pyrargyrite (Ag_3SbS_3), occurs in bunches through the ore and its relations to the other ore clearly prove it to be secondary as it coats cracks and fractures in the ore, especially that which runs high in the argentiferous tetrahedrite. Enrichments by pyrargyrite occur in even the lowest workings of the mine, 1,300 feet below the outcrop, but their number is greatly diminished.

The silver values were contained primarily in the galena and tetrahedrite, chiefly in the latter, but have been redistributed as native silver and pyrargyrite by the action of oxidation and secondary enrichment. In the oxidized ore of the upper workings a very considerable amount of native silver was found.

Banding is not very well developed, as the ore is chiefly massive, irregular, and filled with innumerable horses and fragments of country rock. Where it is at all well marked it is the result of the replacement of sheeted rock rather than of the successive deposition of minerals in open space.

A peculiar brownish-green, massive, fine-grained pyrite, locally known as "brown iron," was found in most of the upper workings of the mine, and is said to have run very high in gold. Four specimens of this ore left for only a few weeks in paper trays completely corroded the paper by the uncombined sulphuric acid, showing the ore to consist of sulphates, sulphuric acid, and pyrite. Probably much of this pyrite is secondary and the high values in gold were undoubtedly the result of secondary concentration. It does not occur below the upper workings of the mine. There is little or no question that the workable values of the Black Crook ore have been entirely produced by oxidation and secondary enrichment, for the lower workings show only primary minerals in small amount and low values. All of the rich minerals are of characteristic secondary types, and extensive exploration has failed completely to show values that justify the continuation of mining operations. The width of many of the stopes may also indicate that secondary minerals have been deposited not only within the area of the original vein minerals, but also in the wall rock, which they may have invaded and replaced in their downward progress, materially increasing the ore masses both in size and in richness of contained values.

The paragenesis of the primary minerals, here as elsewhere, is obscure, but the metallic minerals seem to have been among the earlier formed, and of these the sphalerite and galena seem older than the richer tetrahedrite. The pyrargyrite and native silver are unquestionably secondary.

The ore in the mine, even with the aid of the assay plan kindly furnished by the management, does not show any well-defined localization. If such localization exists, it has somewhat the position shown in figure 24.

To the writers the evidence seems conclusive that further exploration in depth on this vein would fail to reveal any bodies of ore that could meet the expense of development.

The average value of the ore in the Black Crook mine is difficult to determine, as no authentic records of the earlier work could be secured. Much of it was undoubtedly high grade, as shown by the mint report for 1884. During three months' active production in that year the mine yielded 1,277 tons of ore, valued at $124,447, an average of $97.21 per ton. Statements of the superintendent put the average yield of this upper ore at 21 ounces silver and 3 ounces gold. Much very high-grade ore was undoubtedly included in this average, for many small bonanzas and bunches of secondary ruby silver yielded from $200 to $600 per ton and in instances gold values ran as high as 12 to 15 ounces, but the average seems to be fairly close to $100 per ton.

With increase in depth, the value of the ore fell off rapidly, much of it averaging not more than $35 per ton in gold and silver. A careful sampling recently made shows that the average value of the ore now remaining in the mine is much below the bonanza values of its early operation.

The values in the ore show no relation to the Black Crook fault, as the higher values occur more frequently at a distance from it than in its immediate neighborhood. If any connection can be detected at all, it is in the direction of impoverishment and not of enrichment.

SOUTH FORK OF HENSON CREEK.

MORO.

The Moro mine (see fig. 26) is located about 1½ miles in a direction S. 60° E. from Capitol City. The mine is operated by the Hanna Mining & Milling Co. The openings are on the Moro claim on the south side of a small east-west gulch, which heads westward from the South Fork of Henson Creek.

Equipment.—The Moro is connected with Capitol City by a trail about 1½ miles long, over which supplies are brought to the mine. Recently a wire-rope tramway was installed, connecting the mine openings with the mill in the bed of the South Fork of Henson Creek, 1,200 feet below the mine and at the mouth of the gulch in which the mine is located. The ore was formerly shipped to the smelter, but it is now treated in the mill owned and operated by the company. The mill has a capacity of 100 tons per 24 hours, and treats not only the ores mined in the company's own property, but also undertakes custom work from the mines in the neighborhood of Capitol City. In 1906 this mill was much enlarged, and now contains rolls, Huntington mills, concentrating tables, slimers, a system of settling tanks, and a Blake-Morscher static electric machine for handling the zinc concentrates. Power is supplied by the Capitol City Power & Electric Co.

Development.—A fissure vein worked in the mine is opened by three tunnels which run into the hill in a direction about S. 15° W., and connect with the first, second, and third levels, respectively. (See diagram, fig. 27.) The first, or uppermost, level opens directly on the vein and runs into the hill for 500 feet; 200 feet from the mouth it connects with an old shaft, by means of which the vein was first operated. The second level has been driven 71 feet below the first, directly on the vein; it runs into the hill for 570 feet.

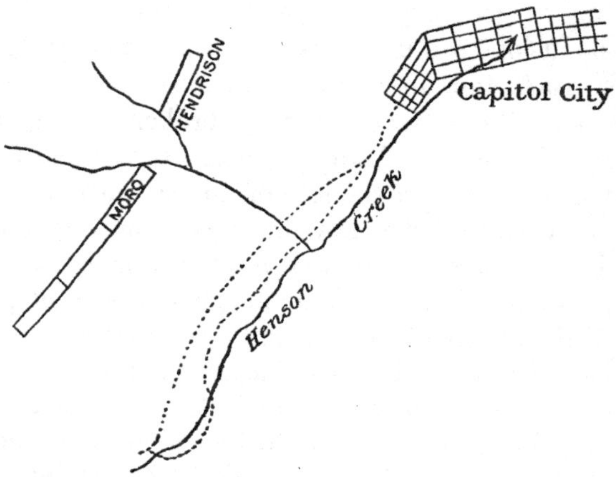

FIGURE 26.—Sketch showing location of Moro and Hendrison claims.

The third, or lowest, level is 126 feet below the second. For the first 75 feet it runs S. 60° W. and cuts through barren rock; it then encounters the vein and runs along it for 700 feet. At a point 200 feet from the breast the third level is connected with the other two levels by an inclined raise on the vein. A number of short drifts have also been run on branch veins on the third level.

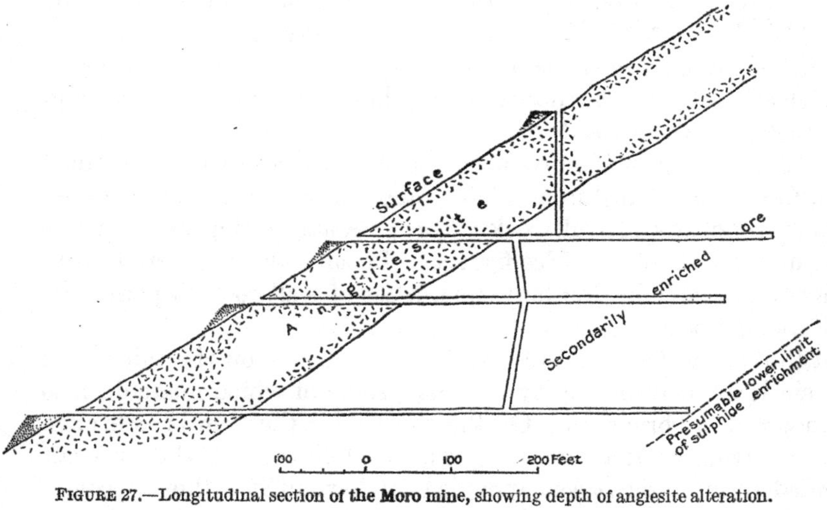

FIGURE 27.—Longitudinal section of the Moro mine, showing depth of anglesite alteration.

Country rock.—The country rock of the region consists of the lower members of the Picayune volcanic group of banded flow breccias. The banding is caused by the movement of the lava previous to consolidation. When fresh the flow lines can be seen only indistinctly and on close examination, but when weathered they become very prominent and may be seen to be caused by innumerable dark inclusions arranged with their longer axes in the direction of flow. The banding is generally horizontal or inclined at a very slight angle. The inclusions are usually darker and finer grained than the matrix of the rock, and undoubtedly represent the quickly chilled coating formed on the surface and later broken and included in the slowly moving lava mass. Many of them are markedly chloritized even at great distances from the site of mineralization. In some places this wall rock shows fewer flow lines, especially in the thicker members, and is characterized by well-marked porphyritic texture, caused by the presence of numerous small phenocrysts of plagioclase feldspar. The distribution of shoots and vein material in the mine is independent of the banding of the lavas, for there is no tendency toward a horizontal variation of mineral material; and it is therefore probable that the layers of andesite have had little or no effect on the ore deposition. Cross mentions (p. 23) a small exposure of granite at the mouth of the gulch near the mill, which indicates that the Picayune volcanic group is nearly cut through here by Henson Creek, and it is possible that this formation lies less than 1,500 feet below the lowest level of the mine.

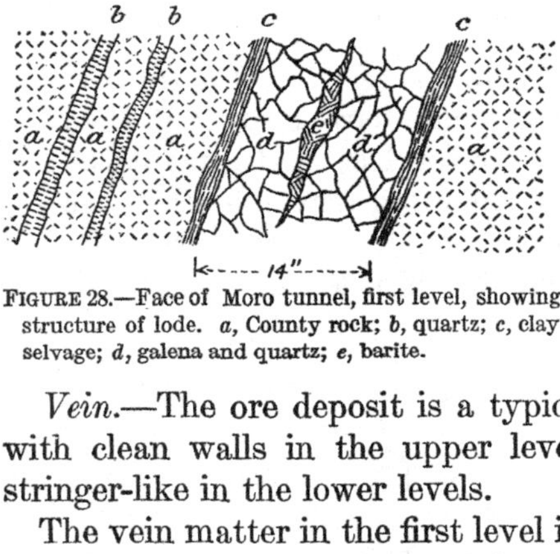

FIGURE 28.—Face of Moro tunnel, first level, showing structure of lode. *a*, County rock; *b*, quartz; *c*, clay selvage; *d*, galena and quartz; *e*, barite.

Vein.—The ore deposit is a typical fissure vein well defined and with clean walls in the upper levels, but becoming irregular and stringer-like in the lower levels.

The vein matter in the first level is contained between well-defined and quite smooth hanging and foot walls, but contains a great many angular fragments of wall rock. Some parallel stringers of quartz occur in the wall rock. (See fig. 28.) Locally the vein splits into stringers and includes too large quantities of wall rock to permit its profitable operation.

On the second level and to a still greater degree on the third level the vein is divided into many stringers, some of which make off into the foot wall as branches. On the third level the vein is so broken up into stringers that it is difficult to follow, and the crosscut intended to intersect it was extended far beyond it without recognition. Beyond the crosscut the vein consists of a series of veinlets en échelon running at about 10° to 15° to the general trend of the lode. (See fig. 30). The dip of the vein at the surface and on the first and second levels is 66° E., but this steepens to 78° on the third level. (See fig. 29.)

In strike the vein is extremely irregular, as may be seen from the plot of the workings (fig. 31). Most of the bends are quite angular as if caused by the intersections of branching fissures.

This vein is undoubtedly the same as that extensively worked on the north side of the gulch. Its total length, as worked on both properties, approximates 2,000 feet.

Ores.—The ore of this mine when unoxidized consists of galena, sphalerite, chalcopyrite, pyrite, and tetrahedrite in a gangue of white glassy crystalline quartz with subordinate barite. The galena is by far the most abundant mineral and constitutes the chief product of the mine. It is generally coarse, with single cubical cleavage masses up to an inch across, but in places it has the usually fine-grained appearance of steel galena. It is everywhere much crushed and often shows twinning due to stress. It is also generally characterized by curved cleavage faces. Many of the stopes consist of this coarse galena, with but little else present, although bunches of sphalerite occur here and there through it. The sphalerite varies from almost lemon yellow to nearly black and where present is usually intermixed with the chalcopyrite in the ore, indicating that it was precipitated simultaneously with that mineral. The argentiferous tetrahedrite is in far less amount than is common in the ores of this region, showing only here and there through the ore. The metallic minerals of the ore are often much shattered and cut by innumerable little veinlets of white glassy quartz, which frequently show a very perfect comb structure and form inclosing crusts around the shattered fragments. (See fig. 32.) Between the combs of quartz are thin bands of a reddish mineral, apparently hematite, which form a delicate tracery that seems to much emphasize the banded nature of these crusts. It is stated by the management that where these fine reddish crusts are present the ore generally contains notable values in gold, and it may be that the increased gold values and reddish

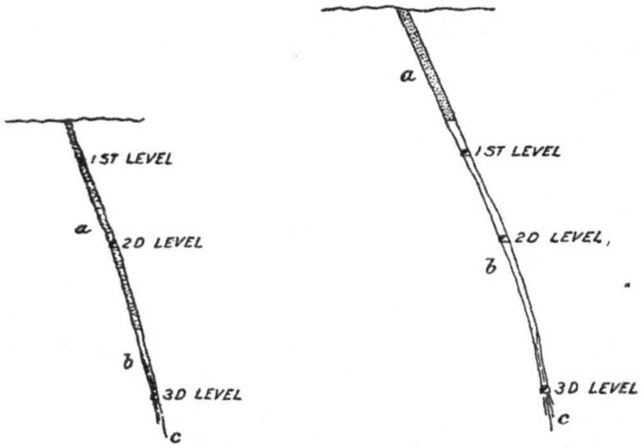

FIGURE 29.—Cross sections of the Moro vein on lines indicated in figure 31. *a*, Anglesite in zone of oxidation; *b*, primary ore; *c*, fraying out of vein on third level.

FIGURE 30.—Manner in which Moro vein crosses fissures diagonally on lower level.

mineral are both due to secondary precipitation. Vugs of considerable size and in large numbers are thus produced in the ore, and these are usually lined with white quartz crystals and some beautiful bladed crystals of barite.

The primary ore of the upper level stopes consisted almost wholly of galena, which carried an average silver content of 10 to 17 ounces. The chalcopyrite and sphalerite were most commonly associated with subordinate amounts of galena. The average yield of the lower-grade ore from the upper levels is stated to have been 10 to 17 ounces silver, 10 to 15 per cent lead, 4.4 per cent copper, and 6 to 15 per cent zinc, giving a total average value of $30 a ton.

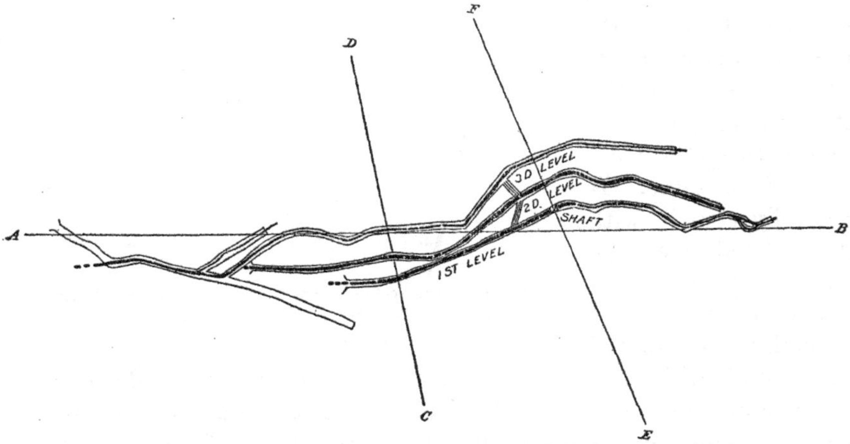

FIGURE 31.—Plan of workings of the Moro mine. (See fig. 29 for cross sections.)

This ore carried only very small and usually insignificant quantities of gold. The primary ore in the third level shows a marked change, containing much more chalcopyrite, pyrite, and sphalerite, and yielding correspondingly higher gold than the ore from the first and second levels. Some of the ore here assayed 0.5 to 3 ounces gold, 10 to 16 ounces silver, 7 to 30 per cent lead, 7 to 10 per cent zinc, and 2 to 4 per cent copper.

The primary vein filling is separated, in the upper levels, from the hanging and foot walls by a well-defined layer of selvage clay one-half inch to 3 inches in thickness, making the ore comparatively easy to mine.

Except that the quartz is of later origin than the metallic minerals the paragenesis can not be clearly made out. It seems probable that the barite is the later of the two gangue minerals as it comes often in bladelike crystals in cavities. Banding is not well developed in the metallic ore of the upper levels as the quartz runs through so much of the metallic ore in little veinlets, but a central vug filled with barite is often present. (See fig. 28.) In the lower levels banding is more prominent (fig. 33).

Secondary alteration.—The surface alteration of this vein is extremely interesting. Oxidation has altered the ore to a dense grayish-white noncrystalline anglesite ($PbSO_4$) for about 200 feet from the surface (see fig. 27); the carbonate, cerusite, is practically unknown in the mine. The anglesite in a quartzose gangue, extends in from the surface for 180 to 200 feet. The line which separates it from the sulphide ore is sharp and follows very closely the contour of the hill. As the sulphide zone is approached little nuclei of original primary galena may be seen in the ore, and the structural arrangement of the galena ore is perfectly preserved in the anglesite.

FIGURE 32.—Ore from the Moro mine, showing certain features of the paragenesis of the minerals. *sp*, Sphalerite; *ch*, chalcopyrite; *g*, galena; *q*, quartz; *h*, hematite.

Below the oxidized anglesite ore considerable enrichment of the ore by secondary copper sulphides is manifested by beautiful blue coatings of covellite on the fracture surfaces of the sphalerite. This sphalerite when fresh has a brilliant vitreous luster and yellowish-brown color, but when broken with a light tap of the hammer it falls to pieces along the innumerable fractures which were formed prior to the secondary deposition. These are so completely coated with covellite that it is at first difficult to distinguish it from that mineral, but when care is taken to break into the mineral itself along cleavage planes not previously opened its true character is apparent.

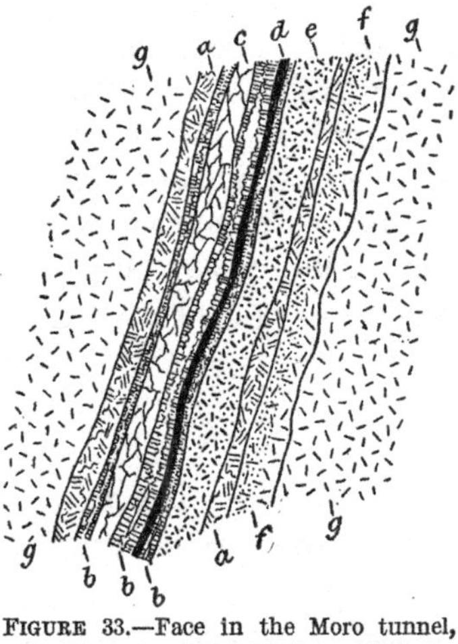

FIGURE 33.—Face in the Moro tunnel, showing structure of lode. *a*, Massive galena and quartz, irregularly mingled; *b*, quartz veinlets with well-developed comb structure; *c*, altered country rock with veinlets of galena and quartz; *d*, narrow band of intensely silicified country rock now altered to aggregate of quartz granules; *e*, blackened silicified country rock; *f*, darkened and partly altered country rock; *g*, andesite flow breccia.

Considerable black sooty chalcocite is also deposited in the ore, but this is confined almost wholly to fractures and cavities in the chalcopyrite, while pyrite is confined to some surfaces of the galena. Some of the sphalerite is also covered along fracture planes with a gray coating which is believed to be galena but which, because of its fine grain, is difficult to identify positively even under the high powers of the microscope. The minerals indicate that the workings of the mine have not yet penetrated below the zone of surface alteration, but already the vein, like others in the district, shows not only marked impoverishment even in the 500 feet of vertical distance so far disclosed, but also every indication of fraying out and disappearing. It holds no greater promise of profitable deep exploration than do the other veins.

HENDRISON CLAIM.

The Hendrison workings are directly north of the Moro on the opposite side of the gulch. The vein is unquestionably a continuation of the Moro and is in every respect similar to it. A tunnel has been run on the vein for 400 feet. The ore carries a higher content of zinc than that portion of the vein worked in the Moro and is also very much more noticeably enriched by pulverulent secondary chalcocite.

PROSPECTIVE GOLD AREA.

Some exploration for gold is being carried on. The Sunshine Lode in Larson Creek is said to show assay values of from $1 to $10 per ton, but as no shipments have been made, it is not possible to say just what value can be placed upon these assays. The lode itself does not look very promising, but has not been developed enough to disclose its true value. The Golden Crown prospect and others in the area are practically undeveloped, and so little can be said of them, except that they are very narrow veins of quartz, showing few metallic minerals and these chiefly pyrite. The T. C. M. Tunnel is being driven to cut at depth several veins which outcrop near Larson Creek, but at present (1908) only reported assays showing total values from $1 to $6 in lead, silver, and gold, can be cited. No shipments have been made. The outcrops do not appear to be very highly mineralized, although greatly decomposed.

BIBLIOGRAPHY.

There are few publications on the Lake City district. The following bibliography includes practically all of the literature bearing directly on the region:

1874 to present. Engineering and Mining Journal, weekly publication. Many notes, usually from correspondents, on the history, development, production, etc., of the Lake City mining district.

1880. Emmons, S. F., Tenth Census, vol. 13, p. 86. Gives a table showing kind of country rock, ore, and gangue minerals, and dip and strike of veins in eight mines in Hinsdale County.

1887. Comstock, T. B., Geology of vein structure of southwestern Colorado. Trans. Am. Inst. Min. Eng., vol. 15, pp. 218–265. Refers to the Lake City region in a very casual manner, giving no definite information on it.

1887, 1888, 1889, 1890, 1891, 1892. Reports of the Director of the Mint. Contain statistics of production of individual mines, 1887–1892; also brief notes on the operations, development, etc., of the various properties.

1889. Schwarz, T. E., Ore deposits of Red Mountain, Ouray County, Colo. Trans. Am. Inst. Min. Eng., vol. 18, p. 140. Mentions the district in connection with several contiguous Colorado areas.

1901–2. Report of commissioner of mines, State of Colorado, pp. 93–97. Gives brief sketch of history, geology, mines, and minerals of Hinsdale County.

1903. Rickard, T. A., Across the San Juan Mountains, pp. 72–77. Contains a description of the Golden Fleece mine, Hinsdale County.

1904. Irving, J. D., Ore deposits in the vicinity of Lake City, Colo. Bull. U. S. Geol. Survey No. 260, pp. 78–84. Gives brief account of the mining district and a short sketch of the history, geology, and ore deposits.

1880–1908. Mineral Resources of the United States. Contains a few references to mines in Hinsdale County, during various periods.

INDEX.

O

Made in the USA
Monee, IL
04 November 2026

67380770R00079